Geographies of Consumption

Juliana Mansvelt

Geographies of Consumption

© Juliana Mansvelt 2005

First published 2005. Reprinted 2008

Apart from any fair dealing for the purposes of research or
private study, or criticism or review, as permitted under
the Copyright, Designs and Patents Act, 1988, this publication
may be reproduced, stored or transmitted in any form, or by
any means, only with the prior permission in writing of the
publishers, or in the case of reprographic reproduction, in
accordance with the terms of licences issued by the
Copyright Licensing Agency. Inquiries concerning reproduction
outside those terms should be sent to the publishers.

SAGE Publications Ltd
1 Oliver's Yard
55 City Road
London EC1Y 1SP

SAGE Publications Inc
2455 Teller Road
Thousand Oaks
California 91320

SAGE Publications India Pvt. Ltd
B 1/I 1 Mohan Cooperative Industrial Area
Mathura Road, New Delhi 110 044
India

SAGE Publications Asia-Pacific Pte Ltd
33 Pekin Street #02-01
Far East Square
Singapore 048763

British Library Cataloguing in Publication Data

A catalogue record for this book is available from the British Library

ISBN 978-0-7619-7429-1 (hbk)
ISBN 978-0-7619-7430-7 (pbk)

Library of Congress Control Number 2004099543

Typeset by C&M Digitals (P) Ltd, Chennai, India
Printed in Great Britain by Cpod, Trowbridge, Wiltshire

For Laura and Hannah

Summary of Contents

1 Geographics of Consumption

2 Histories

3 Spaces

4 Identities

5 Connections

6 Commercial Cultures

7 Moralities

Contents

Preface — xiii
Acknowledgements — xv

1 Geographies of Consumption — 1

Geography and Consumption Matters — 1

Box 1.1 Geographies matter to consumption: eBay – the world at your fingertips — 3

Box 1.2 Consumption matters to geographies: e-trash as a world of waste — 4

Conceptualizing Consumption — 6

Box 1.3 The social life of things: commodified and decommodified moments — 7

Box 1.4 Material culture: why, where and how things come to matter — 9

Geographies of Consumption: Critical Social Science — 10

Spatialities of Consumption — 11

Box 1.5 Alternative economic spaces: local exchange trading schemes — 15

Socialities of Consumption — 16

Box 1.6 Ethical food relations: new socialities, new geographies and ethics — 18

Box 1.7 Horizontal and vertical systems of provision — 19

Subjectivities of Consumption — 20

Box 1.8 Mort: mapping masculinities — 22

Power Matters — 23

Box 1.9 Karl Marx (1818–1883). Producing consumption: commodification and commodity fetishism — 23

Marxism and political economy — 24

Poststructuralism	25
Box 1.10 Foucault: productive power	26
Box 1.11 Power geometries: the social and spatial exercise of power	27
Outline of the Chapters	28
Consumption and Geography Matters	29
Further Reading	30
Notes	31

2 Histories 32

Urbanization, Industrialization and the Emergence of 'Modern Consumption'	32
Box 2.1 When did modern consumption emerge?	33
Nineteenth and Twentieth Century Consumerism: the Spatial and Social Extension of Consumption	35
Box 2.2 Thorstein Veblen (1857–1929). Consuming production: conspicuous consumption and emulation	36
Box 2.3 Advertising and the context of the commodity: commodity racism	38
Emerging 'public' spaces of consumption	39
Box 2.4 'Consuming' women out of place: the New York woman and the middle class shoplifter	40
Fordist production and consumption	42
Alternative geographies	43
Box 2.5 Contesting the hegemony of consumerism: consumer societies	43
Postmodernity and Niche Consumption	44
Box 2.6 Postmodernity and lifestyle shopping	48
'Placing' Transformation in Consumption: Problems of Extrapolation and Interpretation	49
Box 2.7 Capitalism, commodification and consumption in Russia	49

Histories and Chronologies:
the Need for Specificity   52

Further Reading   53

Notes   54

3   Spaces   56

Space, Place and Scale   56

Box 3.1   Lefebvre and Soja: making sense of space   56

Box 3.2   Robert Sack and the consumer's world   58

Spectacular spaces   59

Box 3.3   Theme parks: spectacular spaces
          of consumption   59

Trajectories of Consumption?
Placing Consumption Sites   60

Box 3.4   Benjamin and the arcades project   60

Shopping malls: all-consuming places?   61

Box 3.5   John Goss and the magic of the mall   62

Box 3.6   Reconciling textual and ethnographic approaches   63

Box 3.7   Socialities and subjectivities:
          shopping malls in Turkey   65

Exploring 'alternative' sites of consumption   66

Box 3.8   Alternative spaces of consumption   67

Home: Scaling Public/Private Spheres   69

Box 3.9   Making home and home-making: tools of gender?   70

Geographies of Cyberspace   73

The Internet and cyberspace: inclusions
and exclusions   73

Box 3.10 Consuming cyberspace: linking real
          and virtual worlds   75

Consuming Spaces   77

Further Reading   78

Notes   78

4 Identities — 80

Consuming Identities and the Postmodern Condition — 80

We consume to become who we are? — 81
We consume according to who we are? — 82

Box 4.1 Bourdieu: cultural capital, distinction and identity formation — 83

Body Matters — 84

Box 4.2 Geographies of the body matter: the wardrobe moment — 85

Bodies out of place — 86

Box 4.3 Embodiment and emplacement: landscapes of 'later years' — 87

Linking Embodiment and Emplacement — 89

Performance — 89

Box 4.4 Goffman: performance in frontstage and backstage settings — 89

Performativity — 90

Box 4.5 Butler: performativity, gender and identity — 90

Box 4.6 Body matter: dirt, discourse and second-hand clothing — 92

Placing Consuming Identities: Geographies of Food — 94

Box 4.7 The world on a plate: displacement and identities — 96

Box 4.8 Consuming food: spaces, practices, identities — 97

Commodity and Body Journeys — 99

Further Reading — 100

Notes — 100

5 Connections — 101

Linking Production and Consumption: the Commodity Chain — 101

Box 5.1 Chickens and commodity connections — 101

Global commodity chains 102

Box 5.2 Nike: 'just do it' or 'just stop it'? Consumer activism and global chains 105

Box 5.3 Services and commodity chains: the case of sex tourism 109

Systems of provision 110

Box 5.4 Transnational biographies: the used clothing industry 111

Commodity Circuits 114

Box 5.5 The Story of the Sony Walkman 115

Actor Network Theory 117

Box 5.6 Plants and people: an ANT approach 119

Box 5.7 Linking consumption and production: black boxes and coffee consumption 120

Box 5.8 Shaping motherhood and milk consumption: poststructural political economies 123

Power, Politics and Connectivities 124

Further Reading 125

Notes 125

6 Commercial Cultures 127

Understanding Commercial Cultures 127

Box 6.1 Commercial cultures 128

Music: Exploring the Politics of Commercial Culture 129

Box 6.2 Instore music: the product and performance of consumption 130

Box 6.3 Hip-hop: rap as resistance? 132

Commercial Cultures and Globalization 134

Global homogenization 134

Box 6.4 McDonald's: consuming meanings in sacred and therapeutic landscapes 136

Creolization, hybridity and transnationalism 137

Box 6.5 Transnationalism and commercial cultures 139

Tourism: Consuming Culture as Other? 140

Box 6.6   One hundred per cent pure? Maori as the welcome party of Aotearoa/New Zealand 142

Commercial Cultures as Connections 146

Further Reading 146

Notes 147

7 Moralities 148

Performativity: Seeing, Doing and Becoming Geographies 148

Box 7.1   Consumption knowledges: performing the subject of consumption 149

Box 7.2   Food deserts: deprivation and diet 152

Moral Geographies of Consumption 152

Box 7.3   Commodification, consumer culture and moral economy 153

Box 7.4   Perspectives and praxis: influencing global environmental change? 156

Box 7.5   Questions of sustainability: household consumption practices 158

Box 7.6   Shaping moral geographies: consumer citizens and climate protection 159

Politics and Praxis 160

Box 7.7   The practice of hunger 161

Consuming Geographies 164

Further Reading 165

Notes 165

References 166

Index 187

Preface

Writing In: Fashioning Geographies of Consumption

In reflecting on the diminishing numbers of surveys on consumption, Fine notes that the field has become too vast to cover: 'In short, consumption is a moving, expanding and evasive target, especially in view of the array of the analytical weapons with which it has been assaulted' (2002: 1). I'm glad I read Fine's quote near the end of the process of writing this book! In many ways in writing this book I feel I have been complicit in the production of a 'God trick', implicitly reconstructing myself as an all-seeing, all-knowing, omnipresent purveyor of things consumption. However, this survey of geographies of consumption is necessarily a partial and a situated one, informed not only by what I have read, but also by the multiple subject positions I occupy. I speak not from a placeless 'nowhere' but from somewhere, and that 'somewhere' is located in my embodiment, my emplacement and my practice as a Pakeha New Zealander, geographer grappling with political economy and post-structural perspectives, a Christian, and a mother juggling work and family. Though I can reflect on my own positionality, I can never fully articulate it or know exactly how it affects the research endeavour (Rose, 1997). Nevertheless if reflexive accounts are ones in which 'absences, fallibilities and moments are brought into visibility' (Pratt, 2000: 651) then it remains important to signal something of the silences that have informed my thinking, doing and writing of this book.

Discussing others' research is an act of selective re-representation, a process of 'writing in' meaning (Berg and Mansvelt, 2000) which I am aware is framed largely around the hegemony of Anglo-American geography and the English language. As a New Zealander writing this text I was extremely conscious of the sheer abundance of North American and particularly British literature. While writing from New Zealand might reinforce this Anglo-American bias, I have included a number of New Zealand case studies not only because they are the spaces with which I engage most closely as a geographer, but because narratives from 'the periphery' can challenge and confront hegemonic geographies (Berg and Kearns, 1998). I have nevertheless tried to include a range of case studies reflecting different places, practices and ideas about consumption. Though shopping and retail geographies have been and remain critical to consumption as a subject of geographical research, a concern has been to broaden the focus of the book beyond this.

This book discusses some of the fascinating research being conducted by geographers on consumption histories, spaces, connections, subjects, commercial cultures and moralities. In focusing on some key areas of geographies of consumption I am conscious that advertising and collective and institutional consumption have not been covered in depth, and research on cultural economy, rural spaces, services and geographies of work might also have been accorded more attention.

That consumption is a complex field of social and spatial relations is a recurrent theme of the book. Consumption is constituted in contexts which are connected to other practices, processes and peoples. Much of the research outlined in this text demonstrates the inseparability of production and consumption, the economic and cultural and the symbolic and material. Many of the topics covered in this text – eating, shopping, home-making, tourism, music – may seem mundane, but it is in their taken-for-grantedness that power is most effectively manifest. Though I have drawn primarily on the work of geographers, the work of other social scientists and theorists has been included as it continues to enliven and inform geographies of consumption. Ultimately, this text has endeavoured to show how place matters in how consumption is created, manifested and experienced, and how consumption matters in how socialities, subjectivities and spatialities are constituted in place.

My own personal challenge as a geographer remains to explore consumption in ways which are sensitive to the differences between people, processes and places and which enable people to relate the particular to the general in order to facilitate the meaningful negotiation of the material, symbolic, social and individual aspects of everyday life. Readers may laugh at the ideological, normative and impossible nature of such a task, but endeavouring to make a difference does not, and cannot, sit apart from my intellectual imaginings (informed as they are by the practices, connectivities and representations of others in place: see Gregson and Rose, 2000) or from my practice as a university geographer who teaches and does research. Somewhat ironically then, writing this book has simultaneously been a part of my own consumption of 'the academy' and a project of self-fashioning!

Acknowledgements

My sincerest thanks to Neil, Laura and Hannah for putting up with a wife and mother whose life has for some time revolved around writing 'the book'. Thanks also, Neil, for the endless hours of proofreading! I am also indebted to Dad, Hanny, my family, my friends and my church family for supporting me, and for looking after the girls so I could have extra time to write. I am grateful to my colleagues for their encouragement, and for all the support I received from my Heads of School, previously Professor John Overton and currently Professor Mike Roche. Thanks also to the many people with whom I have had conversations about the book during its various stages, but particularly to Barbara Arnold, Richard Le Heron and Wendy Larner for their input into the final chapter. Special thanks to Olive Harris and Kelly Dombroski for the considerable amount of time taken in compiling the references for this book, and to David Feek for tracking many of them down. Financial assistance from the Massey University Research Fund and a Massey University Women's Research Award enabled me to complete this book. I appreciate so much the advice, assistance and patience of Robert Rojek and David Mainwaring, editors at Sage Publications. To the anonymous reviewer of the book, thank you for your helpful comments. To Rachael Newman who always encouraged and kept me sane while writing this – I am forever grateful!

1

Geographies of Consumption

Buying, using and disposing of commodities connect us to other people and other places in ways which may be beyond our imaginings. Commodities are more than just objects; they are shifting assemblages of social relations, which take place and assume form and meaning in time and space. For many individuals, consumption is both a visible and a pervasive part of everyday life in contemporary society. A trip to a market, a store, a fast food restaurant, the movies, or a local trader may be a taken-for-granted aspect of everyday life for many, but these actions play a critical role in the meaningful creation and expression of place. This book is about geographies of consumption: the ways in which relationships between people, things and places are constituted around the sale, purchase and use of goods and services. It is also about the perspectives geographers have used to interpret these things.

Consumption is fundamental to how geographies are made and experienced in contemporary society. From bodies to nations, cities and homes, through markets and retail outlets, this book examines how consumption occurs, through what processes and in what places. Geographies, as the spatial expression of social and physical processes, are in turn integral to how consumption processes are constituted and articulated. A key theme is the necessity of acknowledging the situatedness of consumption processes, that is, how they take and make meaning as they are created and expressed across space and time. A diverse range of case studies will be used to demonstrate that consumption processes are fluid and contextual, fabricated differentially and unevenly across space. The emphasis of this text, then, is on how geography matters to consumption and how matters of consumption are also geography matters (Clarke et al., 2003: 86).

Geography and Consumption Matters

Consumption has become 'one of the grand narratives of the second half of the twentieth century' (Mort, 2000: 7). The increased visibility of sites of consumption and the proliferation of consumer goods and images have led social commentators to suggest consumption rather than production is now the driving force in contemporary society (Corrigan, 1997). Consumption is so integral to the constitution of contemporary society that it is almost impossible

FIGURE 1.1 Advertising and shopping around London's Piccadilly Circus. Processes and practices of consumption are integral to the constitution of contemporary society

to avoid in capitalist social formations (Bocock, 1993). In fact there are 'few areas of everyday life not affected by or linked to processes and practices of consumption' (Edwards, 2000: 5), and being, working and living in the developed world are dominated by individuals' relationships with consumer goods (Miles, 1998a; Ritzer, 1999) (see Figure 1.1). Consumerism, whereby individuals (both producers and consumers) become enmeshed in the process of acquiring commodities, and formulate their goals in life in relation to the acquisition of commodities, is argued to be so ubiquitous in contemporary societies that it has become 'a way of life' (Miles, 1998a).

The increasing volume, variety and incursion of commodities in everyday life, growing commodification, greater social division and self-reflexivity have been associated with a hypothesized postmodern condition. In the postmodern epoch, commodities are assumed to have a more significant role in mediating social life than was previously the case. Commodities and commodity relations are promoted in popular culture and media as offering liberatory, hedonistic and narcissistic possibilities – being keys to self-realization, happiness and fulfilment (Douglas, 2000). Consumption practices and preferences are also instrumental in identity formation, self-expression and the development of lifestyle cultures built around such things as diet, fashion, music and leisure tastes (Featherstone, 1987: 55).

Landscapes of consumption are said to be increasingly ubiquitous and visible in contemporary society. The increasing incursion of processes of commodification in everyday life is also promoting de-differentiation, or the blurring of previously discrete consumer activities such as shopping and banking services, theme parks and malls, medical centres and shops (Bryman, 1999). In an era in which sites of consumption are increasingly rationalized settings, the theming of places becomes a means of 're-enchantment' – replacing the impersonality and instrumentalism of consumption (Ritzer, 1999). Consumers' practices are changing too, with consumers increasingly servicing more of their own needs (Gershuny and Miles, 1983) and consumption being encountered more and more in dematerialized forms as services and representations and via non-material sites such as the Internet (Slater, 1997). The diffusion of information technology, time–space compression, and the emergence of networking and deepening communication and flows of information through electronic media are viewed as symptomatic of the network society (Castells, 1996) (see Box 1.1).

BOX 1.1 GEOGRAPHIES MATTER TO CONSUMPTION:
EBAY – THE WORLD AT YOUR FINGERTIPS

eBay, the Internet auctioneer site, was set up in the USA by eBay chairman Pierre Omidyar in 1995 and has surpassed Amazon.com, the online bookstore as the world's most popular shopping site (Mesure, 2001). The site is a marketplace for the sales of goods and services with tens of millions of registered users globally. Anything can be put up for sale as long as it is not on the list of 'prohibited, questionable or potentially copyright infringing goods'. Commodities for sale range from antiques, collectibles and crafts to music, real estate, vehicles and industrial and commercial services. As part of eBay's philosophy of connecting people, users were initially engaged in bidding person to person (a process traditionally occurring at spaces such as garage sales and open-trader markets), but now they are also able to make purchases at a fixed price.

Speed of listing is one of the benefits of the sites, with the seller being charged a non-refundable insertion fee and for additional listing options which may help promote the item. eBay notifies the buyer and seller via e-mail at the end of the auction if a bid exceeds the seller's minimum price, and the buyer and seller complete the transaction independently of eBay. eBay does assist with processing credit card payments through its PayPal system, but there may be a number of hidden costs for potential consumers which include shipping costs, goods and services taxes, payment charges and import duties.

For eBay consumers, this virtual world of communication, information and exchange would appear to annihilate space, yet the 'death' of geography has not resulted (Dodge and Kitchin, 2000; see also Chapter 6). Though cyberspace is not a physical space it still retains a geography, and is constituted and understood in relation to material things and places. The creation of cyberspace for exchange of commodities may involve new connections between local and global and the creation of spaces which are simultaneously local and global, but producers and consumers are still located in particular places. Potential consumers, for example, may choose to enter eBay through these place based portals to avoid costs associated with importation, monetary exchange, and delivery. Consumers are even described in terms of online communities – terminology which has been drawn from the linkages of people to territorial spaces. Buyers' choice of opening bids may be framed in relation to estimated costs in non-virtual markets, and the goods and services exchanged virtually generally travel across material sites and nodes (such as postal or freight services) from buyers to sellers. Issues of access to Internet bandwidths, computer and electronic technologies, and constraints on payment options are also related to 'material' geographies. Thus even in cyberspace, a space of hypermobile flows of information, geographies matter to consumption, influencing how it is understood as a meaningful space, how it is experienced, and the forms through which consumption/production relations are articulated.

While in the developed world such changes would suggest that people construct much of their everyday experience in relation to consumer landscapes and their actions as consumers (Sack, 1988), consumption also remains important in the lives of those who do not inhabit these places. Whether understood in terms of abundance or lack, need or desire, the meanings of commodities and consuming practices are not fixed and bounded in place but are fluid, fractured and changing across space in an increasingly interconnected world. Consumption in a globalizing world is thus unevenly constituted, 'characterized by stark inequalities of poverty and wealth, of hunger and malnutrition in some places and superabundance in others, of extravagance and waste amid scarcity and need' (Jackson, 2002c: 283). Such disparities are not unconnected (see Box 1.2): they cannot simply be mapped across developed and developing countries, but are features of the geographies of nation-states (whether classified as developed or developing), regions and localities too.

BOX 1.2 CONSUMPTION MATTERS TO GEOGRAPHIES: E-TRASH AS A WORLD OF WASTE

Consuming commodities can be thought of as a relational process, rather than an isolated act because commodities can be conceptualized as bundles of social relations (Watts, 1999). Commodities are purchased, exchanged and devalued over time and across space. They are given meaning in complex networks of production and consumption through marketized and non-marketized processes. A consequence of the purchase of computer and electronic commodities, for example, is the geographies that arise from the use, disposal, storage and recycling of such material. In the USA replacement of electronic and computer equipment is often easier and cheaper than repair (Silicon Valley Toxics Coalition, 2004). Despite the 'take-back' programmes of some manufacturers, opportunities and incentives for recycling remain low and these commodities often end up as stockpiles of e-waste. In, addition the USA has not signed the 1998 Basel Convention prohibiting the export of hazardous waste from the wealthiest OECD countries to all non-OECD nations and is the primary exporter of e-waste to China. Though the Chinese government has sought to stem the flow of e-waste by vowing to turn away illegal waste from its ports, the trade continues, encouraged by China's cheaper labour costs and less stringent occupational and environmental regulations (Chandran, 2002). In Guangdong Province on China's south coast an estimated 100,000 people scavenge through e-waste. An industry of traders, drivers and sorters has sprung up around the reprocessing of computer, television, mobile phone and other electronics and technology based refuse (Ni and Zhang, 2004). In the Guiyu area of the province, local entrepreneurs buy the trash in bulk and migrant labour is employed to break or burn it into component parts which are then onsold and re-used. Workers are usually unprotected from the damaging effects of lead, cadmium, toner, mercury, barium and beryllium common in hi-tech waste, yet the reprocessing industry provides a livelihood for local inhabitants and has been a critical part of the local economy of villages in Guiyu for the last decade (Ni and Zhang, 2004). Thinking about e-trash connections between the US and China illustrates how consumption in one place may be linked to other places and peoples through complex networks in which production and consumption are mutually constituted. Processes and practices of consumption do matter to how geographies are created and expressed.

Commodification and the symbolic and material practices and spaces of consumption are seen as part of a globalization, remaking landscapes and transforming local cultures. Under 'the march of monoculture' social diversity and difference are said to be erased (Norberg-Hodge, 1999b: 194), with some commentators arguing local cultures, practices and spaces are subsumed under processes of commodification, Americanization (see Figure 1.2), Coca-colonization (Norberg-Hodge, 1999a), McDonaldization (Ritzer, 1993) and Disneyization (Bryman, 1999).

Yet the narrative of consumer culture presented here is a partial one. While commodities and their images, consuming practices and sites of consumption have a greater significance in everyday life for many people, the contemporary world is one in which consumption practices, places, knowledges and meanings have often been expressed in uneven, contradictory and hybrid ways. Geographical research has provided a foil to the generalized descriptions and universal processes implied by notions of contemporary consumer culture. It has provided insights into the ways in which people, places and things may facilitate or negate such processes across space and into how processes of consumption are made, manifested and vary across space. Through studying consumption, geographers have sought to explain how and why economy and culture, the symbolic and the material, collide, demonstrating how the complex

FIGURE 1.2   Eat American in Moscow. The globalization as homogenization thesis (see Chapter 6) suggests places are becoming increasingly similar, citing evidence for this in concepts such as the Americanization of places and consumption objects and practices

meanings and expressions of consumption in place are in turn connected to other spaces and scales, and making a critical contribution to the relationship between society and space.

Conceptualizing Consumption

Consumption can be understood as the *complex sphere of social relations and discourses which centre on the sale, purchase and use of commodities*. Social relations comprise interactions, relationships, encounters and practices between people, places and things, and the outcomes and events which stem from them. Such relations may form structures and institutions (for example, family, state or firm) but they also operate at the level of the individual. Discourses refer to the ideas, knowledge and meanings inscribed in language, material objects and social practices by which people make sense of the world (for example, discourses of ageing which present consumerism as a means of retaining youth). People construct powerful discourses through their actions and interpretations but discourses also regulate social thought and action. Conceptualizing consumption as more than an act of purchase enables one to encapsulate a range of material and symbolic practices and meanings which centre on the sale, choice and selection of goods and services, and their purchase, use, reuse, or resale and disposal.

A commodity is an object of consumption and exchange (Lee, 1993: x). Commodities may be goods (concrete and tangible objects) or services (bungee jumping, dry cleaning), people (even body parts) and ideas (such as intellectual property and patents). This book focuses on processes of consumption which revolve around the purchase, use, reuse and disposal of goods and services by final consumers – individuals who will utilize the goods as ends in themselves (for example, food, apparel, holidays). This is in contrast to consumption which occurs as inputs to the production process (as in a firm's purchase of legal or financial services, or a manufacturer's purchase of cardboard for packaging). The spheres of production and consumption are interdependent: consumption is not simply about the using up of things, but also involves the production of meaning, experience, knowledge or objects – the outcome of which may or may not take the commodity form. Similarly producing objects, experiences, artefacts etc. usually involves consumption of things.

The concept of a commodity is central to understanding consumption. In capitalist societies, commodities exchanged through an economic system assume a use value (the capacity to satisfy a want or need) and an exchange value (the ability to command other commodities in exchange). Lee suggests that the essence of all consumer goods 'can be found in the fact that first and foremost they are commodities' (1993: xi) and that it is this aspect which gives them their distinctive character or soul. While production for profit will give a commodity a distinctive meaning or essence, meanings of objects also arise from their non-commodified moments. Geographers have been influenced by the ideas of Appadurai and Kopytoff, who have highlighted the ways in which commodities are exchanged and circulated outside, alongside and even in contradiction to capitalist relations of market exchange (Box 1.3).

BOX 1.3 THE SOCIAL LIFE OF THINGS: COMMODIFIED AND DECOMMODIFIED MOMENTS

In his book *The Social Life of Things* (1986) anthropologist Arjun Appadurai advocated a 'new' way of viewing the circulation of commodities. Instead of focusing on the forms and functions of exchange, he argued important insights could be gained into the politics of the link between exchange and value by studying the social lives of commodities (1986: 3). Appadurai suggested that understanding the meanings of commodities, their forms, uses and trajectories, could 'illuminate their human and social context' (1986: 5). Igor Kopytoff (1986) in the same volume utilized Appadurai's discussion of the social life of things to show how the value and meaning of things may alter over time. He noted how things became commoditized and decommoditized as they moved in and out of the commodity state. Kopytoff noted, for example, how a person upon capture for the slave trade became a non-person and a potential commodity.

> Once purchased as a slave and reinserted into a particular social setting he/she became rehumanized, but always remained a potential commodity whose value could be realized by sale. The same object may also be viewed by one person as a commodity but by another as something far removed from the sphere of capitalist commodity exchange. Both Appadurai and Kopytoff signalled the importance of the social trajectories and biographies of commodities, a point extended by geographers who have argued that commodities also have spatial lives (Jackson and Thrift, 1995).

Commodification denotes the extension of the commodity form to goods and services previously existing outside the market (Jackson, 1999) but commodities are also objects of cultural symbolic exchange. Anthropologists Douglas and Isherwood suggest commodities are an important means of communication in contemporary society and constitute a 'non-verbal medium for the human creative faculty' both conveying and constituting cultural meaning (1978: 62). They argue 'goods that minister to physical needs, food or drink – are no less carriers of meaning than ballet or poetry' (1978: 72). Thus consumption is 'as much an act of imagination' as it is the using up of things, with spatial and temporal contexts making the link between an object and its meaning (Goss, 1999a: 117). The process of ascribing meaning to objects, and the significance objects have for people, can be thought of in terms of material culture (see Figure 1.3 and Box 1.4).

FIGURE 1.3   Commodities are given meaning as they are incorporated in everyday life. Here using a bottle is a ritual of enchantment which helps turns the commodity (a doll) into 'a baby'

Commodity meanings are constituted through a diverse range of consumption practices including rituals of exchange (choice of purchase and presentation, gift giving), possession (e.g. cleaning, displaying, grooming, discussing) and disinvestment (removing, reusing, relinquishing or discarding) (McCracken, 1988). However, meanings are already inscribed in commodities prior to their consumption and it is often the characteristics of a commodity rather than its utilitarian aspects which are consumed, as in the anticipation of a forthcoming purchase or event (Miles, 1998a).

BOX 1.4 MATERIAL CULTURE: WHY, WHERE AND HOW THINGS COME TO MATTER

Material culture can be thought of as the tangible creations (such as artefacts, buildings, crafts, décor, art, tools, weapons, furniture) that society makes, uses and shares. Understanding material culture involves asking questions about how, why and where things come to matter, and how in turn things come to influence the construction of meaningful and power laden social relationships. In his 1987 book *Material Culture and Mass Consumption* Daniel Miller suggests that contemporary societies are increasingly established around the presence of objects, with objects, society and culture being mutually constituted. Like Appadurai and Kopytoff (see Box 1.3) Miller dismissed a concept of goods and services in which value exists only in their commodified form. Refuting a view of consumption as intrinsically negative, fetishist or alienating Miller also draws on Simmel's (1978 [1907]) work on objective culture[1] to emphasize the positive and meaningful ways in which consumption can contribute to expression of self and relationships with others in everyday life (Ritzer et al., 2000). As a consequence of this, spaces become more than just place settings for consumption of objects and services. For example, Miller (2001b) describes how he is 'haunted' by his inability to redecorate his house in a style in which it was constructed – one he aspires to, but cannot achieve. He is also let down by the material environment by which he presents himself to others. In this sense commodities as objects acquire agency; as Miller says, 'Where we can't possess we are in danger of being possessed' (2001b: 120). Material culture also has consequences which may not be an expression of one's agency or which may be unintended (such as when people read differing notions of taste into one's choice of home furnishings) (2001b: 109–11). Miller's ideas have been influential in geography, particularly with regards to ethnographic studies of consumption sites and the socialities associated with them.

Consumption, as the previous discussion has demonstrated, is not a singular, monolithic or static phenomenon but a complex sphere of relations and discourses which are actively (but not always knowingly) assembled, reproduced and expressed in diverse ways in place. Practices and objects of consumption and their meanings also 'travel', occupying different social and spatial 'moments' and being transformed over time and across place. These spatialized trajectories, the biographies and geographies of commodities, are themselves consumed and produced (Watts, 1999). Geographers have provided critical insights into the ways this occurs and into the landscapes and processes which result.

Geographies of Consumption: Critical Social Science

Attention to consumption by social scientists appears to have emerged as a response to the development of a consumer society, commodification, and the predominance of consumption in structuring everyday life in the contemporary world (Edwards, 2000). Given the disparities between First and Third Worlds it is perhaps not surprising that academic interest in consumption, both supportive and critical of the notion of a consumer society, burgeoned in the 1990s (Jackson, 2002c). This is also the case for geography.[2] The publication of Sack's *Place, Modernity, and the Consumer's World* in 1992 was critical in establishing links between consuming practices and geographies (see Chapter 3). Such was interest in consumption that by 1995 Gregson was able to entitle her review article for the journal *Progress in Human Geography* 'And now it's all consumption?'. The 1990s also marked the publication of a number of significant edited social science collections on consumption (Burrows and Marsh, 1992; Edgell et al., 1996; Miller, 1995). These volumes sought to highlight the significance of consumption in the structuring of social life and to present consumption as a legitimate and worthwhile field of academic enterprise. In seeking to demonstrate the impossibility of thinking of consumption in a 'simple, one dimensional way' (Warde, 1992: 28) these publications (including geographers' Wrigley and Lowe's 1996 *Retailing, Consumption and Capital,* and sociologist John Urry's 1995 *Consuming Places*) were instrumental in demonstrating that place, space and scale make a difference to how consumption is manifested and experienced. The significance of leisure and tourism activity in contemporary social change (see Figure 1.4) has also led to a growing volume of geographical work on leisure/tourism practices and spaces with insights for consumption (Aitchison et al., 2000; Crouch, 1999). A defining feature of geographical consumption research over the last two decades has been the insights which have emerged through research collaborations and dialogues across disciplinary boundaries (reflected in publications such as Jackson et al.'s 2000 *Commercial Cultures;* Miller et al., 1998; and Clarke et al.'s 2003 *The Consumption Reader*).

Geographers have nevertheless provided key theoretical and empirical contributions to social scientific consumption literature.[3] They have not only filled the 'gaps on the map of consumer society' (Crewe and Lowe, 1995) by examining sites and landscapes of consumption but, as mentioned previously, produced theoretically informed research which emphasizes the detailed, complex and differential expression of consumption in place and the connections linking spaces.

FIGURE 1.4   Darling Harbour, a former Sydney dockside area now redeveloped into a site of restaurants, exhibition centres, tourist attractions and shops. Spectacular spaces of retail, leisure and consumption initially formed the focus of much geographical work

Geographic research has been at the forefront of efforts to overcome divisions between economy and culture, production and consumption (Sayer, 2001).

Geographies of consumption encompass a wide diversity of subjects: leisure, tourism, work, shopping, information technology, retailing, advertising; urban, rural, industrial and agricultural geographies; and studies of gender, ageing, ethnicity and sexuality. Table 1.1 presents a summary of some of the broad themes of geographical research, but these should not be seen as mutually exclusive or static categories. Geographies of consumption are marked by attention to a diverse range of studies underpinned by different theoretical traditions and perspectives which together provide insight into spatialities, socialities and subjectivities associated with consumption.

Spatialities of Consumption

Geographers have been concerned to explore *spatialities* of consumption. This includes a consideration not only of the places in which consumption is perhaps most visibly and/or actively created (tourist spaces, mega-malls), but of the sites that may be less visible (bodies as surfaces of inscription, home spaces) and the ways in which places and spaces are connected and made meaningful through consumption.

TABLE 1.1 *Themes in geographies of consumption*

Mode	Emphasis	Focus of study
Spatialities ↑↓	Visible, often spectacular spaces of consumption	Examples: spaces of festival and carnival, theme parks, department stores, malls, landscapes of advertising and branding, gentrified spaces, tourist spaces, urban and rural consumption spaces • consumption sites as spaces of representation and representational spaces (landscapes as text) • the historical and contemporary production of space • commodification and commodity fetishism (hidden relations of production) • political economies and structures of race, class, gender • place marketing and promotion • cultural economy
	Mundane, alternative and ephemeral spaces of consumption	Examples: markets, car boots, local exchange trading schemes, Internet, home, workspaces • geographies of everyday life • spaces as relational and social spaces • emphasis on consumers, and blurring of production/consumption roles and relationships • continuities and differences between historical and contemporary formations of consumption in place • construction of value • performativity
	Social–spatial connections	Examples: transnationalism, displacement, global commodity chains, systems of provision, circuit and actor network approaches, commercial and commodity cultures
Socialities ↑↓		• the spatial and social constitution of consumption–production relationships • geographies and biographies of commodities (the social lives of commodities) • social–spatial relationships between producers and consumers • power geometries • consumer activism and politics • state and consumption relationships (discursive and material shaping of consumption, collective consumption) • regulation

TABLE 1.1 *(Continued)*

Mode	Emphasis	Focus of study
		• globalization
• spatial imaginaries
• cultural economies
• retail geographies
• imbrication of material/symbolic culture/economy/production/consumption
	Social relationships in consumption	Examples: shopping, purchasing, using, discarding, eating, leisure practices, home provisioning and renovation
		• ethnographic and social geographies of consumption
• emphasis on situated and social (rather than individual) context of consumption
• emerging forms of sociality (e.g. via new technologies)
• cultural politics
• moralities of consumption
• relationality of spaces, scales, processes and practices of consumption
• material culture (role of commodities in social life)
• decommoditization
Subjectivities	People as consuming subjects, bodies and identities	Examples: studies of self-fashioning and mundane provisioning, role of consumer discourses and consumption practice as part of subject formation, advertising discourses, bodies out of place, consumers as state subjects, embodiment and emplacement of identities, consumer narratives
		• emphasis on poststructuralist perspectives and social construction of consuming subjects in place
• shaping of consuming subjects
• consumption and identity formation
• geographies of consumption and gender, race, class, sexuality
• geographies and discourses of the body
• consumer agency
• moral ascription
• social exclusion
• performativity
• non-representational and embodied practice
• cultural politics

Increasing attention has turned to the relational nature of space and scale, with places actively constituted in relation to other places, existing as fluid and changing entities rather than as tightly bounded containers which simply provide the setting for social processes. Thus the places and spaces and scales across which consumption and commodity meanings, expressions and experiences are assembled and transformed in themselves constitute particular spatialities. Spatialities are thus concerned with sites of consumption, but also with how people, things and processes (such as commodity knowledges) travel – transforming, reproducing, contesting, creating and disassembling particular constellations of social-spatial relations in place as they move.

Early work by consumption geographers made geographies of consumption visible, exploring their expression in sites of consumption (Jackson and Thrift, 1995). Aside from retail spaces geographers have examined a wide variety of sites, from cities to rural 'post-productive' spaces, heritage and tourist spaces, theme parks, festival and carnival settings, gentrified districts, cyberspace, leisure spaces, home and body spaces. There has also been considerable research on how institutional agents (such as local authorities, chambers of commerce) are promoting cities and rural spaces and developing places as sites of consumption for locals, inward investors and tourists, but emphasis has been on the production of these sites rather than on their consumption *per se* (Ateljevic, 2000). Research on historic places and practices of consumption has also offered insights into contemporary consumption practices.

Research on consuming places has become more theoretically informed with studies intended to explicate the material and symbolic construction of sites of consumption. In the early 1990s significant attention was devoted to the ideological construction and significance of consumption space (Jackson and Thrift, 1995). This involved semiotic approaches and techniques of reading landscape to explore the social construction and power embedded in built environments. Much of this work was influenced by Marxian traditions and by the writings of Baudrillard on signification and alienation, with spectacular spaces of consumption viewed as sites of commodity fetishism, illusory places of pleasure, leisure, hyper-reality and simulated 'elsewhereness' (Hopkins, 1991).

From the mid 1990s ethnographic studies emerged, in part a response to concerns that readings of consumer landscapes were resulting in a view of consumers as passive subjects reproducing discourses and structures of consumption framed by producers (owners, designers, marketers, managers and advertisers) (Gregson, 1995). Much of this research centred on what was first known as 'alternative spaces of consumption' such as car boot fairs, second-hand and discount stores, high streets and markets – sites of mundane and everyday provisioning for many people (Crewe and Gregson, 1998; Gregson and Crewe 1997a; 1997b). These studies revealed the ways in which places and practices of consumption are relational, for example, how malls and high streets are understood in relation to each other as 'safe' or 'risky' racialized shopping environments

(Miller et al., 1998). In this process, emphasis has tended to broaden from the study of sites to the examination of how consuming practices and meanings are constituted across space. Looking beyond processes of capitalist accumulation, the importance of structures and discourses other than those connected with the accumulation of capital, or with gender, sexuality, age, race, class, (dis)ability and postcolonialism, has also been highlighted. While retail and shopping geographies still appear to dominate geographical research on consumption, other spaces which have previously been deemed marginal (spatially, socially or economically, such as LETS: see Box 1.5) and spaces of mobility and ephemerality (such as airports, journeys and cyberspace) provide exciting new avenues for consumption research.

BOX 1.5 ALTERNATIVE ECONOMIC SPACES: LOCAL EXCHANGE TRADING SCHEMES

A consideration of spatialities of consumption involves thinking not only about places, but about how consuming processes are constituted across particular spaces. Local exchange trading schemes or systems (LETS) provide an example of an alternative space of exchange which has endured despite the hegemony of capitalism. Since the 1980s these locally based systems of exchange have provided a form of community currency. Originating in Canada, they now operate in Britain, Australia, New Zealand and Western Europe and in Latin America, Asia and Africa. Participants in LETS are both consumers and producers, contributing and 'purchasing' skills and services (for example, child minding, plumbing, decorating and complementary therapies) and products and produce (such as handicrafts, baking, office products and leisure goods) (Pacione, 1997). The producer of a service or artefact is paid in non-commodified fictitious currency credits which are then exchanged as the participants consume other skills, services or commodities. Real money does not change hands; rather a credit/debit account is maintained. No interest is paid and credit is available as participants do not need to have a positive currency balance to consume.

 The value of the trade produced and the economic significance of LETS at national levels may be relatively small, but what matters is their value to members (Seyfang, 2001). LETS have reconfigured dominant consumption, production and exchange processes in order to tackle issues of social and financial exclusion, and have generally been associated with 'green' ideologies, empowerment and community building. By expanding the social network of those who may be marginalized in society, and through recognizing the value and contribution of work in the informal sector, LETS provide not just a 'space' for the expression of alternative values (Seyfang, 2001: 993) but a form of life politics by which these values might be operationalized. Research with LETS participants in the UK suggests that sociality of consumption in such networks can outweigh their material or economic significance (Purdue et al., 1997: 657–9; Williams, 1996).

> Local exchange trading schemes promote direct links between consumers and producers and rupture any associations of the former with passivity and the latter with activity. It is the consumer's productive role in consuming which helps another member to earn credits. Though consumer societies and LETS have the potential to provide a form of resistance to the hegemony of capitalist consumption and production, they do not exist in isolation from it. Many LETS members, for example, were reluctant to allow their accounts to go into debit because the concept of indebtedness was transferred from the cash economy (Aldridge and Patterson, 2002). Ironically, the potential of LETS for 'community renewal' and 'social inclusion' may also be limited as LETS become increasingly regulated by state institutions, and begin to reflect some of the gender and class characteristics of the wider community.

Increasingly geographers are exploring how people, entities and things are caught up and shaped within spatial systems and networks connected with particular geometries of power (see Box 1.11). Researchers have, for example, highlighted the spatial connections in production/consumption systems and the ways in which exploitative commodity relations might be unveiled (Hartwick, 1998). They have also considered how the politics of home consumption might be linked to other scales, processes and institutions in powerful ways (Leslie and Reimer, 2003). The 'pursuit of commodity stories' – research on the social and spatial lives of commodities across space – has become an increasingly prominent feature of geographical narratives on consumption (Bridge and Smith, 2003). This research explores not only what moves across space, but how meanings of consuming and of commodities become removed or displaced from their original contexts (Cook et al. 1999), demonstrating that spatialities of consumption are deeply intertwined with socialities, a concept to which we turn next.

Geographical research on sites of consumption, and the spatialities which characterize consumption practices in and across places, has been valuable in emphasizing how processes of production are deeply embedded in consumption and vice versa. This research has demonstrated how spatial relations are powerfully constituted through the practice of consumption and has signalled the importance of space and place in the construction, representation and reproduction of processes of consumption.

Socialities of Consumption

Another focus of geographic research has been *socialities* of consumption – the connections, relationships and social interactions between people. Though consumption has often been portrayed as a self-centred, narcissistic and individualist enterprise, geographers (particularly those working within ethnographic traditions) have demonstrated the sociality and commensality associated with consumption. They have also noted the ways in which individuated practices are formed in relation to others, whether they are located near (as in family members or work colleagues) or are imagined and distant others (as in racialized understandings of food practices, or 'distant' recipients of clothing donated to charities). Socialities are not static; the relations are in a constant state of flux. The emergence of new forms of socialities (for example, the creation of Internet chatrooms or indigenous tourism ventures) has provided insights into the meaning and value of commodities, the significance and nature of consuming practices and spaces, the politics of consumption and the interdependence of consumption and production (see, for example, Dodge and Kitchin, 2000; Holloway and Valentine, 2001a; Wrigley et al., 2002).

Research on the socialities connected with consumption 'at home' has been fruitful in the breaking down of dichotomies (such as between work and leisure, public and private, production and consumption) and in thinking how relationships and discourses extend to other scales. Feminist geographies exploring patriarchal and sexualized discourses in space have demonstrated how relatively mundane activities such as eating and preparing food and dressing the body are politicized (Banim et al., 2001; Valentine, 1999a; 1999b; 1999c). Detailed ethnographic research has also demonstrated how commodity practices are productive of powerful discourses which define boundaries and relationships between people in everyday life and how these are linked to material geographies (for example, in differences between how shopping is talked about and practised: Gregson et al., 2002a). Much of this research has been underpinned by the concept of 'cultural politics', concerning how meanings are constructed and negotiated and how relations of dominance and subordination are established, defined and resisted (Jackson, 2000: 140–1).

In addition, ethnographic research exploring the socialities of second-hand cycles of consumption and on the Internet has challenged constructions of value and commodity exchange as market-like, rational and profitable. This research has provided insights into how values are made and transferred and shift in and out of commoditized market relations in situated contexts (the social lives of commodities). The worked at, pleasured, leisured and creative aspects of the production and consumption of objects nevertheless remain under-researched; more research is needed on what people do with commodities, how they repair, restore, value and devalue them (Crewe, 2000; though see Hetherington, 2004). An area in which limited research has been done is into the socialities and ethics associated with practices of acquisition and use (see Box 1.6) and particularly with regard to material need, poverty and lack (Cloke and Widdowfield, 2000) and the super rich (Beaverstock et al., 2004).

> **BOX 1.6 ETHICAL FOOD RELATIONS:**
> **NEW SOCIALITIES, NEW GEOGRAPHIES AND ETHICS**
>
> Lewis Holloway's research on Internet enterprises illustrates how consumption of commodities is underpinned by geographical and ethical socialities. The 'My Veggie Patch' online service presented customers in London with the opportunity of having vegetables grown for them in Suffolk. Consumers could decide what they wanted to grow and how their vegetables were to be grown, but the work of gardening was done for them and their produce was delivered direct to their door (2002: 73). A second site enabled consumers to adopt a sheep in the mountainous Abruzzo region in Italy, giving the client 'direct contact with the origins of what they eat' (2002: 74). Using an actor network perspective (see Chapter 5), Holloway argued both sites 'bring home' distanced food production to consumers and allow producers to provide a service to consumers without traditional intermediaries such as food retailers or distributors. However, in order to perform these roles, new associations and assemblages of people, entities and place must be established (for example, connections between customers, the Internet, adopted animals, vegetables, money etc.). Holloway's research demonstrates how the formations of socialities and spatialities are inseparable, and he highlights the way in which these new encounters between distanced things also involve the creation of particular sets of ethical relations (for example the notion of participatory care of the things involved in food production, or engaging with the risks of production such as disease).

There has been a significant amount of research on more formal characterizations of socialities and how these are manifested across space, that is, as frameworks and systems of social–spatial relationships. Research on consumption networks and systems, commodity chains and regulatory frameworks tended, at least initially, to be focused on institutional socialities arising from linear or vertical chains of connection in political–economic contexts (see Box 1.7). This research has made and continues to make a valuable contribution to debates about consumer sovereignty, power and the interdependence of production and consumption, for example in the linking of consumers with moral/ethical company practices (Johns and Vural, 2000; Silvey, 2002). Consumer protests about issues of globalization, sustainability, free trade and the power of transnational corporations have drawn attention to the influence of states, non-governmental organizations and transnational corporations in mediating social relations of production and consumption (Klein, 2000). Literature on the connections between retailing, regulation and consumption, for example, has also been critical in highlighting the spatial organization of retail capital and explaining the processes, structures and mechanisms which connect production and consumption (Wrigley and Lowe, 1996; 2002).

> **BOX 1.7 HORIZONTAL AND VERTICAL SYSTEMS OF PROVISION**
>
> The work of Ben Fine and Ellen Leopold (1993; Fine, 2002) on systems of commodity provision has also been influential in work on consumption, in which geographers have sought to explore the interdependence of consumption and production, and the socialities and spatialities associated with this. Horizontal approaches are concerned with factors which apply across society and consumption as a whole (Fine, 1993: 599), such as changes in the advertising sector, retailing or distribution. Fine and Leopold based their systems of provision approach instead upon vertical chains of connection, exploring the multiple processes (and sites) of production, distribution, marketing and consumption and the material culture which surrounds the production and consumption of commodities. They note how vertical systems alter depending on the commodity and the structures and histories in which they are embedded (see Chapter 4), thus allowing for recognition of spatial and temporal contextuality of production–consumption processes. Vertical approaches have been important in conceptualization of commodity chains, and have provided valuable insights into how commodities are created and transformed across space and how social practices and knowledges may be articulated through different spatial contexts. However, geographers Paul Glennie and Nigel Thrift (1993) have critiqued the reductionism of this distinction between horizontal and vertical approaches – a distinction that circuit and network based approaches to conceptualizing consumption/production socialities and spatialities are suggested to have overcome (see Chapter 5).

Actor network (Whatmore and Thorne, 1997) and circuit approaches (du Gay et al., 1997) have emerged to address some of the concerns about the linearity and productionist bias of socialities framed from a political-economic perspective. These approaches also offer potential for thinking about 'space' itself, for understanding space as 'multiply inhabited and characterized by complex networks, circuits and flows' (Crang et al., 2003: 441). However, there remains a need for studies which explore multiple sites along networks, chains or circuits (Jackson and Thrift, 1995) and for more work on the spaces, socialities and subjectivities associated with consuming services (as dematerialized or ephemeral commodities).

A concern has been to move beyond accounts which privilege 'economically driven explanations to the detriment of culturally sensitive accounts' (Crewe, 2000: 276). Work on cultural economies which looks at the interdependence of the two spheres has burgeoned in the last five years (du Gay and Pryke, 2002; Scott, 2000), with studies of commodity and commercial cultures (Crang et al., 2003; Jackson, 2002a) trying to examine the imbrication of the two spheres rather than the inflection of each in the other (as in cultural industries, or the production of consumer culture). Work on commercial cultures has focused on the particular socialities associated with the constitution and transformation of the contexts and narratives surrounding commodities to understand how people and entities are caught up in transnational networks (Dwyer and Jackson, 2003). As with consumption practices, a recognition of the ways in which such networks are shaped in relation to moral and ethical concerns and environmental understandings is gradually emerging as a research agenda (Hobson, 2003; Wilk, 2002).

Though geographical research points to a view of socialities as institutional and collective interactions as well as individual social relationships, work on collective consumption appears to have waned in recent years (Fine, 2002). Nevertheless, research has continued on the role of the state and the local state as regulator and 'producer' of goods and services either directly (such as in provision of housing) or indirectly (as in tourism and leisure as a tool of local economic development and place marketing strategies). New Zealand geographers, for example, have explored the links between regulatory frameworks, commodification and the changing role of the state in documenting the emergence of consumerism in healthcare in a deregulated and neoliberal New Zealand economy (Barnett and Kearns, 1996; Kearns and Barnett, 1997) and the commercialization of tertiary education in New Zealand and Canada (Berg and Roche, 1997). However, considerable scope remains to examine how social practices and relations of consumption are regulated, legitimated, maintained and/or contested by people. Little is known, for example, about how discourses of 'consumer choice' are shaped by state and other agents in place, or how they are negotiated through consumption (Pawson, 1996).

Subjectivities of Consumption

The third concept, which has figured compellingly in the geographical literature, is the notion of *subjectivities*. Eschewing categorizations and typologies of consumers, geographers have examined how consuming subjects are made and performed through personal and collective acts, discourses, relationships and imaginings.

Different perspectives in human geography construct different narratives of subjects: for example, Marxian understandings of subject formation occur in relation to one's structural position, while humanist understandings locate subject formation in the autonomous capacities of human agents. The concept of the 'subject' utilized in this text draws from poststructural theory which emphasizes the ways in which people, individuals, bodies and identities are socially constructed in relation to others who are separate from 'self'.[4] Much geographical consumption literature has also explored subjectivity through the lens of poststructuralism, arguing subject formation is an expression of power relations established through the disciplining operation of discourses. Discourses operate through processes of exclusion and boundary making which define self and others. However, while subjectivities may be established through discourses they are not determined by them, and a key contribution of geographical studies has been to highlight their active constitution in specific spaces and the ways in which subjects are prone to moral ascription.

Geographers have emphasized the material and symbolic constitution of consumption, contributing to understandings of the ways in which people meaningfully experience and incorporate commodities in their life worlds. Much research had centred on the practice of everyday life (de Certeau, 1984), exploring how consumers are actively involved in using commodities and practices in ways that might differ from those intended by producers of such spaces (as in the formation of oppositional subcultures: Hebdige, 1979). An important finding of this research is that much commodity use and meaning are not anxiety ridden or identity based, but centred upon material, social and familial relationships and rationales (Valentine, 1999c).

Research on consumer subjectivities has extended beyond a consideration of issues of representation and textual readings of advertising, media and place based landscapes to explore the role of the consuming subject and the meanings people give to consumption practices and commodities (Stevenson et al., 2000). This work has destabilized the notion of 'the universal and passive consumer', instead focusing on the agency of consumers, examining the 'work' and skills employed in choosing, purchasing and using commodities, and the interpretations, sociality, tensions, meanings, emotions and knowledges inscribed in consumption practices (Crewe, 2001; Williams et al., 2001). Another significant area of research has been on the role of consumption in the place based constitution of social identities, problematizing the traditional separation between production and consumption identities (McDowell and Court, 1994).

Geographical research on consumption and identity formation has contested notions of homogeneous consumer cultures based around specific patterns of consumption, revealing that a critical part of the production of space involves how consumption is made meaningful through processes of embodiment, emplacement and performance (Skelton and Valentine, 1998). Poststructural research has emphasized bodies as surfaces of inscription, as powerful spaces, as sites no less critical to consumption than the 'spectacular' landscapes of consumption which initially dominated geographical attention. Work on embodiment has demonstrated how bodies are disciplined and objectified through consumption, for example by marking old or 'fat' bodies as out of place (Gamman, 2000; Gibson, 2000).

The way in which consumption practices intersect with relations and discourses of gender, sexuality, age, race and class has enabled geographers to understand how people (though primarily adults) are constructed as consuming subjects (see, for example, Box 1.8). Cultural politics has been demonstrated to be important in how subjects ascribe meaning to processes and practices of consumption. In exploring the subjectivities associated with consumption of second-hand clothing, Gregson et al. (2000) noted how consumers negotiated bodily associations and potential histories of commodities, making them susceptible to bodily discourses in ways that other personalized commodities (books, electronic gear, whiteware) may not be. The emphasis on the role of bodily practices, movements, senses and habits (which are not necessarily subject to discourse) by which human beings meaningfully engage with and perform consumption (Thrift, 2000c) has also challenged the visual bias of discursive and representational understandings of consuming subjects. Consequently, through researching subjectivities, geographers have made important insights into both material and representational practices of consumers, into the often blurred distinction between consumers and producers, and into the ways in which identities and practices surrounding consumption are performed, reproduced and manifested in place.

BOX 1.8 MORT: MAPPING MASCULINITIES

In his studies of place and masculinity, Frank Mort has endeavoured to explore the power relations which are an integral part of how consumption is manifested, expressed, experienced, invested with knowledge and resisted (Mort, 1998). In his study of the 'Archaeologies of city life' (1995) Mort examined the ways in which masculine subjectivities were framed in discourses and material practices which surrounded historical and spatial change in London's Soho district in the 1980s. The social and sexual identities of homosexual men in Soho were shaped by practices and sites of consumption. Mort described how media and cultural entrepreneurs in the 1980s drew on the district's lexicon as a Bohemian site of sexual and cultural dissidence, youth styles, and artistic and literary culture which 'had been laid down at different historical moments' (1995: 577) to frame a particular rhetoric of style. This rhetoric was located in the newly emerging professional subjectivities and the commercial transformation of the area with the associated development of shopping, leisure and entertainment facilities. However, such constructions of social space were partial, ignoring the significance of gender and marginalizing people and practices centred on other forms of city life. In the 1990s a number of carnivals and gay commercial ventures contributed to a renaissance of the district as a space for gay consumption and homosocial relations. Mort notes how both 'formations were predominantly masculine, though they evoked quite different interpretations of geography and identity' (1995: 581). Mort went on to demonstrate how the arrangement of Soho's consumer culture drew these diverse communities into adjacent social spaces. His research demonstrates how place plays a powerful rather than a passive role in the formation of subjectivities, and how consumption practices can create diverse discourses and productions of place.

Spatialities, socialities and subjectivities of consumption are not mutually exclusive. Together they constituted geographies of consumption – geographies which are about the complex relationships between social and spatial relations, the ways in which people, material and symbolic practices, entities and things are connected, performed, transformed and expressed as they are created and move across space. Implicit in the discussion of spatialities, socialities and subjectivities is the question of how power is shaped, wielded and manifested through material and discursive practices in place.

Power Matters

In researching spatialities, socialities, and subjectivities geographers have brought critical insights to the operation of power, examining the taken-for-grantedness of consumption processes and practices and contributing to understanding how people, identities and bodies are disciplined and differentiated in specific contexts. While political, social, economic and cultural geographers have drawn on a wide range of social theorists to understand consumption (from Baudrillard's systems of objects and signs to Bakhtin's historical analysis of carnivalesque, from Bourdieu's cultural capital to Goffman's and Butler's performativity),[5] debates about how power and the politics of consumption might be conceptualized appear to have been dominated by two intellectual traditions: the employment of Marxian theory in political economy approaches, and the insights offered by poststructuralism.[6]

Earlier research on consumption tended to position it as a consequence of economic production imperatives (Jackson and Thrift, 1995). Prior to the 1980s a significant amount of attention in economic and industrial geographies was directed at how commodities were provided (rather than consumed *per se*), and a predominant strand in this was the use of political economy approaches which drew on the writings of Karl Marx.

BOX 1.9 KARL MARX (1818–1883) PRODUCING CONSUMPTION: COMMODIFICATION AND COMMODITY FETISHISM

In the development of his historical materialist analysis of societal change, Marx interpreted the commodity in terms of the production process. Marx saw consumption as necessary to realize the exchange value of commodities and consequently essential to the continued survival of the capitalist mode of production through the accumulation of surplus value. In the *Grundrisse* Marx stated that though production was the predominant moment, it was determined 'by other moments' (1973: 96). He believed production, distribution, exchange and consumption were inseparable as 'members of a totality', with production and consumption each creating the other in completing itself (Harvey, 1982: 80).

> Marx's concept of 'commodity fetishism' has been utilized by geographers to understand the way in which commodities (as bundles of social relations) may obscure exploitative relations of commodity production. This is a process whereby 'the value of the commodity assumes the guise of a value independent of human determination and thus appears to reside as an intrinsic property of the commodity itself' (Lee, 1993: 14). So while exploitative labour relations inherent in capitalist relations of production are reflected in the process of market exchange, the social and symbolic nature of the commodity form matters too as the product of labour becomes commodities which are both social and sensuous things.
>
> Within Marxian theory, social relationships between people are reconstituted as relationships between things. Processes of commodification mean people are obliged to become consumers and purchase products they and others have made in the workplace (Miles, 1998a: 17). Commodification results in exchange values supplanting use values over time, as commodities are produced for market exchange rather than direct utilitarian value *per se*. As a consequence people are alienated or estranged from the products of their labour, and spatial and social division between production and consumption results (Lee, 1993). Ironically consumption performs a palliative role, offering both recompense and reward for processes of alienation, but never really achieving either because the individual is entwined in multiple processes of false consciousness (Edwards, 2000).

Marxism and political economy

Marxian theorization and the political economy approaches derived from it focus on the economic forces that gives rise to consumption (Box 1.9). Political economy approaches have emphasized the role of actors in their institutional settings, often paying particular attention to how state and economy are articulated through historical and material structures to produce societal transformation. The application of Antonio Gramsci's (1971) concept of hegemony, a refinement of Marx's concept of dominant ideology, has enabled scholars to consider how power operates ideologically through production and consumption. Hegemonic social relations are ones in which the capacity to control other groups does not arise out of a totalitarian exercise of power, but operates through institutions (such as the state, media, advertisers, retailers) in ways which may be subtle, hidden and taken for granted, so that the effect of power is often accepted and reproduced by citizens unknowingly. Marx's concept of commodity fetishism was also taken further in the work of Adorno and Horkheimer (1944) who suggest that mass consumption (and the culture industry in particular) is part of the ideological maintenance of capitalist society (Edwards, 2000). In their rather pessimistic view, the culture industry exists as a form of propaganda and manipulation which 'perpetually cheats its consumers of what it perpetually promises' (1944: 11).

In developing the concept of commodification, Marx provided a mechanism for societal change. His work recognized the interdependence of production and consumption and the redemptive qualities of consumption for the 'alienated' individual; it also pointed to the symbolic value of commodities and their ideological function. Geographers have used Marx's ideas to think about notions of freedom and constraint in consumption, to explore hidden commodity relations in commodity chains and systems of provision, and to examine the hegemonic construction of landscapes of consumption (e.g. Goss, 1999a). Marx's ideas of commodity fetishism have also been taken up in the writings of Cook and Crang (1996), who suggest contemporary capitalist societies are now subject to a double fetish whereby the imaginative geographies associated with commodities become as significant as the hidden commodity relations in things in constituting commodity meanings (see also Castree, 2001).

Marxian approaches have been criticized for examining consumption as a consequence of historical, spatial and social changes in production, as a pleasure seeking but repressive pursuit in which consumers are passively engaged and actively exploited (see Shammas, 1993, and Fine, 2002, for an opposing view). The primacy of class and labour relations in the structuring of production and consumption relations can also render other structures of social differentiation (sexuality, race, gender, age) less visible.

Nevertheless Marxian inspired work has underscored the complex and contradictory nature of consumption: its utilitarian and ideological aspects, its material and symbolic manifestations, and its role in the reproduction of social relations which may be simultaneously alienating and redemptive, socially divisive and socially cohesive. The impact of poststructural perspectives and approaches emanating from the 'new cultural geographies' seems however to have led to renewed scrutiny of previously taken-for-granted meanings of concepts such as 'production', 'consumption', and 'consumerism' and hegemonic constructions inherent in particular conceptualizations of society and space.

Poststructuralism

Poststructuralism is not a unified theory but a number of approaches drawing on semiotics, cultural theory and psychoanalysis which emphasize not only the material forms through which the world is structured, organized and manifested, but the way it is represented in relation to other things (Ward, 1997). The role of language as a sign system is seen as critically important to how meaning is produced and expressed in the social world. Poststructuralism views power laden relationships as internal rather than external to other types of relationship (such as economic relationships), rejecting notions of stable, coherent foundations to meaning and the idea of a central or universal truth. Meaning and identity are relational constructs, constantly being created out of difference (e.g. between self and other, or between texts) – acts of creation which may centre on what is suppressed, absent or excluded as much as what is present. Discourses define appropriate ways of seeing and acting, placing limits on what can and can't be said.

Some poststructuralists employ 'deconstruction' to uncover the hidden assumptions that are embedded in discourses as taken-for-granted ways of approaching the social world. The ideas of Foucault (see Box 1.10) have also been influential in poststructuralist approaches.

BOX 1.10 FOUCAULT: PRODUCTIVE POWER

Foucault sees power manifest not so much in people, but through a distribution of bodies, surface, lights and gazes which operate via an arrangement whose internal mechanisms produce the relations in which individuals are caught up (Foucault, 1979). The concept of surveillance is used to describe this automatic functioning of power, the governing at a distance which Foucault believes is symptomatic of contemporary society. Foucault (1979) used Bentham's concept of a panopticon (a tower from which everything can be seen without the observer ever being seen) to describe how surveillance operates to produce power in everyday life. Through his historical studies of prisons, mental institutions and sexuality, Foucault demonstrated how people become subjects of power, assuming responsibility for its constraints and engaging in power plays which may reflect and reproduce the discipline of the panoptic gaze (Ward, 1997). Thus Foucault's conceptualization of power was as productive rather than simply repressive, with people being subjects of and subject to powerful discourses. Power was not a thing wielded but a process which could be taken over by institutions (states, hospitals, schools, families) and apparatuses of governance (police, administration, familial relations). Power is expressed in discourses which establish claims to truth, defining what counts as legitimate or illegitimate statements, narratives and practices. Discourses consequently operate as systems of regulation in which forms of knowledge and power are produced.

Human subjects exercise power by reproducing and resisting discourses, while simultaneously being subject to them. Subjectivities are created out of this difference between self and other, with socially constructed identities (such as old, young, male and female) being fluid and contextual, formed and changing in relation to the discursive regimes that are part of how time and place contexts are constituted.

A focus on the discursive power created within the sphere of consumption has been augmented by geographers interested in the cultural politics of consumption. Cultural politics thus recognizes the ways in which cultural constructions which are part of everyday life may perpetuate inequalities of power (Jackson, 2000a: 141). This work has been associated with research on subjectivity, identity formation and representation but has increasingly explored the imbrication of cultural and material practices. In contrast to traditional political economy approaches that understand situated behaviour as being underpinned by structural economic factors, poststructuralists would see behaviours, practices, strategies and techniques of power produced by the interconnections between discursive and material realms (W.N. Pritchard, 2000). Consequently studies of cultural politics of consumption have shifted beyond a consideration of the representation of people, places and things to highlight how power is effected to create social, material and moral geographies of inclusion and exclusion in both contemporary and historical contexts.

In seeking to explore consumption in its own right rather than in relation to its role as a particular manifestation of production, Fine (2002) has suggested this has resulted in too much emphasis on the cultural sphere and a relative neglect of the economic and material. However, studies on commercial and commodity cultures (Chapter 7), poststructuralist political economies (Chapter 5) and ethnographic social geographies (Chapters 3 and 4) have all explored how power is manifested through discursive and material contexts.

Though power is conceptualized differently in the two approaches I have outlined, they both continue to provide valuable insights into how consumption geographies are made and expressed in place. Geographers have emphasized how the effects of power are not free-floating but are 'placed' and relational, constructed in social-spatial networks which comprise consumption geographies. Power is spatially produced (see Box 1.11).

BOX 1.11 POWER GEOMETRIES: THE SOCIAL AND SPATIAL EXERCISE OF POWER

Doreen Massey's (1999) concept of 'power geometries' is useful for thinking about how space, place, production and consumption are connected. Massey proposes that power is exercised through all scales and levels, and that its 'geometry' must be understood in relation to how different social groups and individuals are placed in distinct ways in relation to time–space flows and interconnections (Massey, 1993). Difference cannot just be conceived in terms of a variance but must also be seen in terms of relational power, with people limiting or enabling the capacity of other groups to participate in consumption on the same terms. People, knowledge and things are situated in relation to flows and interconnections (for example, transportation, financial flows, communications, knowledge and social transactions). Places are viewed as articulated and hybrid moments in networks of social relations, with 'the spatial as a product of power-filled social relations' (1999: 41). The concept of power geometry thus encapsulates the unevenness within which individuals and groups operate and are positioned in relation to these flows and influences, and the multiple trajectories they create in power–knowledge systems. The concept of 'power geometries' provides an important reminder that consumption spatialities, socialities and subjectivities do not occur in a vacuum, but are constituted and transformed through space.

While approaches to consumption informed by different theoretical frameworks may not be reconciled easily, each offers differing insights into consumption processes. The interconnection of culture, economy, production, consumption, the material and the symbolic is a recurrent theme in this book. Similarly, I have not endeavoured to separate socialities, subjectivities and spatialities, but instead have discussed them as interconnected aspects of the topics which form chapters of the book. The book chapters examine substantive areas of geographical research: histories, spaces, identities, connections, commercial cultures, moralities.

Outline of the Chapters

The first chapter has examined how consumption has been conceptualized and has outlined some of the key contributions of geographies of consumption to social science. In researching socialities, subjectivities and spatialities, and the power that underpins them, geographers have produced significant insights into how consumption matters to geography, and how geographies matter to consumption.

Chapter 2, on histories, examines historical geographies of consumption, exploring the debates surrounding contemporary consumption and the theorized existence of a postmodern condition. The discussion highlights the impossibility of conceiving of consumption as a singular and undifferentiated process in time and space. Examining continuities and differences across place and time provides a way of interrogating chronologies and highlights the specificity of historical geographies of consumption. It also enables reflection on the insight they hold for understanding contemporary consumption practices and places.

Chapter 3, on spaces, explores sites and spatialities of consumption. Places are viewed as shifting and relational assemblages of social and spatial relations. Consumption is seen as vital rather than incidental to how geographies are created and experienced and how consumers and practices are embodied and emplaced, politicized and performed. This chapter explores both formal and informal sites of contemporary consumption to illustrate how spaces of consumption are both consumed and produced, noting that geographical perspectives on consumption in place have the potential to challenge traditional notions of place, space and scale.

The idea that identity formation is the primary reason for commodity purchase and related practices is critiqued in Chapter 4 on identities. The chapter emphasizes the grounding of consumption in the actions, experiences and imaginings of consuming subjects and the ways in which subjectivities are constituted in particular social and spatial contexts. Issues of corporeality and consumption are discussed with reference to processes of embodiment and emplacement. Drawing on the concept of performativity, the chapter then demonstrates how bodily activities and consumption practices operate productively even as they are enmeshed in relations of power. Ultimately this chapter critiques perspectives which suggest consumption is a superficial, individual and passive exercise.

Modes of connection form the substantive focus of Chapter 5. This chapter endeavours to examine three main approaches to linking consumption and production across spaces. Challenging the notion of consumption as a bounded sphere, and essentialist notions of consumers and spaces which derive from this, the chapter examines commodity chains, circuits approaches and actor networks as metaphors for thinking about the movement of commodities, and the spatialities and socialities through which these occur.

Commercial cultures are examined through a series of case studies in Chapter 6. The chapter emphasizes the inseparability of cultural and economic processes in the construction of spatialities, subjectivities and socialities of consumption. Debates about globalization are examined and the chapter advocates approaches which centre on the situatedness rather than homogeneity or universality of cultural/economic change. The chapter demonstrates how perspectives on hybridity and transnationalism enable geographers not only to explore connections between commodities and subjects across space, but to understand how space itself is characterized and made meaningful through complex assemblages of people and things.

The final chapter of the book, on moralities, brings closure to the text by exploring the possibilities and limitations implied by the performative and moral nature of consumption geographies. It suggests different approaches are continually becoming in their capacities to do different sorts of work, to make different subjects and objects and to effect different power geometries in turn. Moral geographies are implicit in the subjectivities, spatialities and socialities of consumption and in the practices and products of geographical research. Moralities of consumption which position consumption as intrinsically negative are critiqued. Following Cloke (2002), I argue for a politics of consumption which is not just sensitive to others but is for them – for a space of politics which becomes a space of deliberation and practice with transformative potential.

Consumption and Geography Matter

Consumption matters to geography. Consumption is fundamental to how geographies are made and experienced in contemporary society. From bodies to nation-states, globally and locally, via real and virtual space, consumption is constituted through places and spaces. Consumption is significant as a place-making process as it is an integral (rather than incidental) part of everyday life, whether or not commodities are scarce or in abundance. Geographies are, in turn, integral to matters of consumption. Geographies of consumption are unevenly expressed in space and make a difference to how consumption practices, entities and experiences are constituted.

By using a variety of theoretically informed methodologies, geographers have provided critical insights into the creation, expression, nature and diversity of consumption practices and meanings in place. They have reflected on the politics of consumption to illustrate how institutions, identities, relationships and practices are produced, reproduced and represented powerfully across space. Geographers have researched interconnections between production and consumption, economy and culture, the material and the symbolic – challenging the construction of these things as dichotomous and equivalent categories. Geographical work on consumption has also provided profound insights into how concepts such as space, value, scale and identity might be conceptualized in specific contexts. In exploring the connections between people, place, practices and entities, geographers have begun to understand how commodities, practices, experiences and knowledges are created and how they travel and translate across space and time.

Such is the volume of work on consumption in contemporary human geography that it is impossible in the space of this text to provide discussion of all geographical research on consumption. The remaining chapters of the text attempt to provide a sample of the diversity of approaches to consumption and the kinds of topics geographers have examined. Consumption is a powerful and pervasive process in contemporary society but it is not placeless: geographies do matter! I hope that readers of this book will gain some insights into how and why this is so, and the possibilities implied for critical understandings of society and space.

FURTHER READING

Bridge, G. and Smith, A. (2003) 'Guest editorial. Intimate encounters: culture – economy – commodity', *Environment and Planning D: Society and Space,* 21: 257–68.

Clarke, D.B., Doel, M.A. and Housiaux, K.M.L. (2003) 'Introduction to Part Two: Geography', in D.B. Clarke, M.A. Doel and K.M.L. Housiaux (eds), *The Consumption Reader.* London: Routledge. pp. 80–6.

Crewe, L. (2000) 'Progress reports. Geographies of retailing and consumption', *Progress in Human Geography,* 24 (2): 275–91.

Crewe, L. (2001) 'Progress reports. The besieged body: geographies of retailing and consumption', *Progress in Human Geography,* 25 (4): 629–41.

Crewe, L. (2003) 'Progress reports. Geographies of retailing and consumption: markets in motion', *Progress in Human Geography,* 27 (3): 352–62.

Goss, J. (1999a) 'Consumption', in P. Cloke, P. Crang and M. Goodwin (eds), *Introducing Human Geographies.* London: Arnold. pp. 114–21.

Jackson, P. and Thrift, N. (1995) 'Geographies of consumption', in D. Miller (ed.), *Acknowledging Consumption: a Review of New Studies.* London: Routledge. pp. 204–37.

Jackson, P., Lowe, M., Miller, D. and Mort, F. (2000) 'Introduction: transcending dualisms', in P. Jackson, M. Lowe, D. Miller and F. Mort (eds), *Commercial Cultures: Economies, Practices, Spaces.* Oxford: Berg.

NOTES

1 George Simmel's (1978 [1907]) discussions of modernity touched upon the role of consumption in society. He saw a tragedy of culture, emerging through exponential growth of the objective culture of material and immaterial commodities, and the subjective culture of people's capacity to use and control these objects successfully. Despite this tragedy Simmel saw commodities as performing liberatory functions as media for the expression of identity and freedom (Ritzer et al., 2000).

2 A search on GEOBASE, a primary electronic database for geography, reveals that the 1990s and early 2000s were the most prolific period in terms of writing on consumption and geography. In the 13 years prior to 1990 the database lists 373 records for consumption and geography; there are 1922 records in the 13 years from 1990. While this may be a function of changes in the breadth of database cataloguing, the difference would appear to be substantial.

3 For comprehensive reviews of the geographies of consumption I would draw your attention to Gregson (1995), Jackson and Thrift (1995) and Crewe (2000; 2001; 2003), whose papers I have utilized here. For more general discussion of consumption research in the social sciences see Fine (2002) and Miller et al. (1998).

4 Subjects exist through the relationship between the world (the other, objects that are distant to the subject) and themselves (Rodaway, 1995). A subject can be thought of as a reflexive and corporeal entity constituted materially and discursively, who possesses the capacity to act but whose actions are both productive of and subject to the operation of power. For example, Marxian understandings have subject formation occurring in relation to one's structural position, while humanist understandings locate subject formation in the autonomous capacities of human agents.

5 The ideas of these theorists will be discussed at other points in the book.

6 This is not to deny feminist approaches, which have contributed significantly to an understanding of gender differences, patriarchy, unfairly structured gender relations, and geographies of difference and diversity. Rather it is to suggest they too have been influenced by Marxism as exemplary radical and socialist theories, and in more recent times by poststructuralist and postcolonial theories.

2

Histories

The visibility of consumption landscapes in contemporary societies could easily lead one to believe that consumerism is a recent phenomenon. This chapter challenges this idea by examining a chronology of consumption which begins with the emergence of modern consumption prior to the industrial revolution and which is currently expressed in a 'postmodern epoch'. Geographers have illustrated the ways in which many consumption practices are contiguous across time, and have looked beyond universal trajectories of consumption to examine the diverse and often ambivalent ways in which consumption processes have been expressed in place (Figure 2.1).

Urbanization, Industrialization and the Emergence of 'Modern Consumption'

Consumption has been an intrinsic and necessary part of society since humans first created, exchanged and used objects. Forms of consumerism existed in Asia and Africa long before Western forms emerged. Despite the long existence of consumerism, qualitative change over time has occurred. The term 'modern consumption' is generally applied to the period when consumption assumed dominance in the structuring and maintenance of everyday life for a majority of individuals, rather than when consumption beyond meeting material needs first appeared (Ackerman, 1997: 109). The period most commonly associated with the emergence of modern consumption or the 'consumer revolution' is the eighteenth century (Campbell, 1987; McKendrick et al., 1982). The industrial revolution, beginning in Britain, which occurred around 1750–1850 was integral to this, establishing a capitalist system based on the formation of industrial capital and the separation of production from consumption through the mass (rather than artisan) production of commodities (primarily for final consumption). The employment of factory based waged labour and the consequent availability and accessibility of factory produced commodities contributed to the rise of consumerism.

However, 'The transformation of modernity itself into a commodity, of its experiences and thrills into a ticketed spectacle, of its domination of nature

FIGURE 2.1 The Roman Baths in Bath, England. Geographers have begun to explore historical contiguities and contrasts in consumption processes. During the Roman Empire public baths such as these were significant in the conduct of business as well as pleasure. In contemporary society such 'heritage' sites have become commodified spaces of tourism. Commercial cultures are consequently not new, but may be manifested and expressed in very different ways over time

into domestic comfort, of its knowledges into exotic costume, and of the commodity into the goal of modernity' was occurring well in advance of the arrival of mass production and consumption (Slater, 1997: 14–15). Geographers Paul Glennie and Nigel Thrift (1992) have suggested why this is so, arguing that consumption was a central feature of artisan and proto-industrial production which existed prior to factory industrialization (see Box 2.1).

BOX 2.1 WHEN DID MODERN CONSUMPTION EMERGE?

Paul Glennie and Nigel Thrift argue that consumption practices were significant in the everyday life of Britons prior to the industrial revolution. Urbanization in England from 1650 onwards (but especially from 1700) contributed to physical and relative distance between producers and consumers and the consequent creation of consuming as a significant social and economic practice (Glennie and Thrift, 1992). They suggest new

> consumer practices emerged in new socialities which developed from the close associations and interactions of urban living. In urban settings, new knowledges about commodities, practices and experiences of consuming emerged. The creation of new discourses based around 'novelty', for example, influenced how goods were consumed and interpreted and (re)produced selectively by different consumers. Discourses of novelty were applied to both familiar (textiles, furniture, metalware) and new objects (tea, coffee, chocolate, ceramics). Mass consumption was also not simply a late eighteenth century phenonemen. Tobacco and sugar products were consumed extensively by the end of the seventeenth century, and tea early in the eighteenth century (Shammas, 1993). Consumption was actually a contributory factor to processes of industrialization as capitalist enterprises developed to meet the rising demand of the elite for luxuries (Ackerman, 1997: 111) but also to provide commodities for final consumption by 'other classes'.
>
> So what effect did the industrial revolution have? In Glennie and Thrift's (1992) opinion, factory produced commodities of the late eighteenth century were incorporated into existing (and evolving) consumption discourses. Industrial production of commodities influenced the price, availability and market condition of commodities and resulted in the rapid alteration of social and economic relationships between producers, consumers, distributors, retailers and purchasers. Modern consumption was marked by greater access to a wider range of commodities in the general population, 'rampant consumer behavior' and an acceptance of consumerist attitudes which extended into spheres of politics, leisure and sport, and commodity production (McKendrick et al., 1982: 11–14).

By the mid eighteenth century a consumer society existed in Britain, France, the Low Countries and parts of Germany and Italy (Stearns, 2001) but the emergence of 'modern consumption' was uneven both in Europe and across the globe. Even within eighteenth century England, the hegemony of consumption was not fully established, with pre-capitalist traditions still shaping popular understandings of exchange. Grain, for example, was available locally at markets on specified days at prices which were publicly agreed as 'morally just' and with the poor of the village often given preferential treatment (Ackerman, 1997). Relationships between mass production, urbanization and consumption were also differentiated commodity by commodity (Fine, 1993).

From the eighteenth century the role of women in procuring commodities for the domestic sphere was accentuated. The cult of domesticity which arose in the nineteenth and twentieth centuries constructed women (particularly of the middle classes) as primary purchasers of commodities for home and family (see Boxes 2.2 and 3.9). Consumerism gained a new respectability through this bourgeoisie domesticity, liberated from discourses associated with the ostentatious leisure pursuits of an elite class and the excessive and risky practices of the working class (Slater, 1997). The rise of consumerism has also been linked to Romanticism, a set of ideas and values based around individualism, emotion, aestheticism, morality and physical beauty as means of expressing one's individual essence. Romanticism was manifest in a range of recreation related activities and commodities such as theatre, horse racing, the novel, poetry and fashion (Campbell, 1987).[1]

Modern consumption also reflected and reinforced Enlightenment thinking which celebrated material progress, rationality and productivity (Stearns, 2001). New technologies and associated cultural change, the production and dissemination of prints, increases in literacy and the advent of newspapers helped to promote a 'modern' worldview and influenced the geographical spread of consumerism (Cressy, 1993). According to Slater (1997) modernity was inextricably bound with consumer culture as the dominant mode of cultural reproduction in Western society, with attitudes shifting 'from prudence and restraint to display, extravagance and the valuing of "newness"' (Edwards, 2000: 34).

Though Enlightenment thinking was founded on secular rather than religious discourses, these have also influenced consumption processes and practices. The Islamic faith, for example, was historically more favourably disposed to consumerism than Buddhism and Christianity, which initially inhibited practices of getting, spending and possessing (Stearns, 2001).[2] Thus attitudes to, and practices of consumption have not occurred simply or solely in relation to particular modes of production, but have been constituted in contexts in which the relationship between civic values, concepts of individual and social identity, political economic traditions, and social-economic constellations are dynamic and changing.

Nineteenth and Twentieth Century Consumerism: the Spatial and Social Extension of Consumption

It is assumed that the availability of new factory produced commodities created a demand which could only be satisfied by the continuing and further industrial production of goods (Cross, 1993). The nineteenth and earlier part of the twentieth centuries saw the extensive spread of mass markets and mass consumption into North America, and the penetration of capitalist relations of production and consumption into colonial 'outposts'. Processes of commodification ensured objects acquired use and exchange values in a market system. More and more economies became organized according to these relations, and were tied into circuits of consumption and production which contributed to the development of a capitalist world system and unevenly developed geographies at various scales (Wallerstein, 1983).

Prior to the establishment of a 'consumer society' mass produced commodities were often unavailable, or were too expensive for the general population.

Over time, goods and services no longer existed as the province of the wealthy but were accessible to those on lower incomes. It is proposed that those in the lower echelons of society sought to emulate the lifestyle choices of the upper and middle classes through fashion, travel and food tastes (McKendrick et al., 1982). However, the concept of emulation has been subject to considerable critique (see Box 2.2). Style became important for all classes, with demand and desire for commodities stimulated by advertising and designs created to serve mass markets (Glennie and Thrift, 1992). Children also became consumers as 'objects' of parental consumption practices and targets for advertisers (Plumb, 1982: 286).

BOX 2.2 THORSTEIN VEBLEN (1857–1929). CONSUMING PRODUCTION: CONSPICUOUS CONSUMPTION AND EMULATION

Thorstein Veblen's *The Theory of the Leisure Class* (1975 [1899]) acknowledged the role of culture in explanations of consumption. Veblen believed economic understandings were unable to explain the complexities of modern life and argued for a concept of the consumer which was active rather than passive (Mason, 1998). Critiquing late nineteenth century North American middle class society, Veblen suggested the wealthy engaged in conspicuous displays and practices of consumption in order to indicate their power and status and difference to others. As a society's level of affluence changed, a middle or 'leisure class' would emerge centred upon consumerist practices (Edwards, 2000). Leisure in Veblen's schema was non-productive consumption of time. Consumerism and pleasure separated the leisure class from the world of work and were an integral part of the economic and material (re)production of class. The historical struggle for existence thus became one of 'keeping up appearances' (Mason, 1998), in which emulation of the consumption practices of the leisure classes by the lower classes in society was seen by Veblen as a principal mechanism for historical changes in consumption. McKendrick et al., for example, state that 'the rich led the way' in developing practices of modern consumption, indulging in an orgy of spending, an 'orgy which was imitated in the lower echelons of society through processes of emulation and class competition' (1982: 10).

Edwards (2000: 26–7) asserts Veblen's theory of the leisure class has made three significant contributions to understanding consumption. First, it accentuated the symbolic and vicarious nature (rather than the utilitarian or exchange based value) of commodities.[3] Second, it demonstrated that consumer practices can be a source of both social cohesion and social division. Third, in its discussion of women, it unintentionally highlighted the oppression of women and the gendering of consumption.[4] The emphasis on lavish spending for extravagant rather than functional reasons has to some extent reinforced associations of consumption with wastefulness, idleness and non-productivity (as in the 'morals' ascribed to the New York woman in Box 2.4).

> Veblen's concept of emulation as a key mechanism in social change has been criticized. Empirical evidence suggests, for example, that choices and practices of consumption are constituted by a complex range of material and cultural factors and that emulative tendencies are not necessarily a key rationale. For example, the working class drew on middle class notions of propriety selectively rather than emulatively (Glennie and Thrift, 1992) and most ordinary consumers purchasing homeware or apparel in the eighteenth century did not actually mimic aristocratic styles (Stearns, 2001). Emulative tendencies also tend to position consumers as unthinking individuals obediently following the latest class based fad and fashion, an idea which was critiqued in Chapter 1.
>
> Nevertheless, Veblen's work has been important in highlighting the intersection of culture and economics in establishing consumption as a significant social practice for individuals and groups, one which emphasized the role of human agency rather than structures *per se* (a notion which, though infrequently cited, appears to underpin social and ethnographic geographies of consumption). In addition, Veblen's notion of consumption as a symbolic practice is a theme taken up by geographers who have explored landscapes of consumption as representational spaces.

During the latter part of the nineteenth century, throughout the twentieth century and into the twenty-first century, consumption has become an increasingly visible aspect of daily life in many nations. This has occurred through the extension and creation of a wide range of consumer practices (shopping, eating out, going to the movies: see Figure 2.2), consumer goods (cars, household appliances, communication devices, computers) and consumer services (everything from dry cleaning to package holidays) and through the commodification of things which previously existed outside the capitalist relations of exchange (such as sport, media images, and even individual subjects and bodies).

Numerous processes have been associated with twentieth century change in consumption: the mass consumption of mass commodities, increased separation of consumption from production (primarily through the separation of home from work), commodification, an intensification of distribution, marketing and advertising of commodities, and a growing connection of consumption practices and identity formation. Many of these processes are not 'new' but have been accentuated, intensifying qualitatively and spatially (Glennie and Thrift, 1992). Marketing, branding and advertising, for example, had their precedents in the emergence of modern consumption in the seventeenth and eighteenth centuries (McKendrick et al., 1982) but in the nineteenth century became integral rather than incidental to the social construction of commodities (see Box 2.3). Cross (1993: 164) believes that during this period the consumption of goods became an intrinsic part of leisure (which he defines as free time) and came to be a means of self-actualization.[5]

FIGURE 2.2   It is suggested consumption activities in their various forms have become an increasingly visible aspect of daily life in many nations over the last two centuries. In recent decades in Palmerston North, New Zealand, café culture has become increasingly popular

---

BOX 2.3 ADVERTISING AND THE CONTEXT OF THE COMMODITY: COMMODITY RACISM

A danger of chronologies of consumption is the subsumption of structures and discourses of sexuality, race, age, gender, religion, health and disability, and imperialism which have intersected with consumption change to produce varying geographies and experiences for consumers. McClintock in her book *Imperial Leather* (1995) notes for example how early advertising employed commodity racism to promote the desirable features of products, racializing domesticity, disciplining 'race' and fetishizing cultures of empire, 'turning imperial *time* into consumer *space*' (1995: 216). In a chapter entitled 'Soft-Soaping Empire' she noted how in the mid nineteenth century soap was more obtainable, promoted by technological changes in production, sources of oil for soap from the colonies and the burgeoning group of middle class consumers. Victorian soap advertising from the mid nineteenth century began to confound images of private and public space, bringing the private world of hygiene and domesticity into the public realm, and bringing scenes of the empire into the world of the home. The fetishizing of soap through advertising promised 'spiritual salvation and regeneration through commodity consumption' (1995: 211), but it did so by relying on racialized notions of

> empire, with images of washing away 'blackness' and 'colour' (the Pears soap campaign in the mid 1880s depicted a black child's skin being lightened upon contact with the soap), the transformation of nature into culture, and the metamorphosing of colonial subjects into consumers. McClintock's discussion of how commodity culture was constituted in relation to particular imperial, gendered and racialized discourses which had their basis in social-spatial relationships and places has been taken up by geographers. Domosh (2003) explores how the gendered and civilizing racial discourses of the Heinz Corporation were important to the establishment of its commercial empire comprising 'colonies' of consumers, while Hollander's (2003) writing on contemporary 'supermarket narratives' shows how contemporary constructions of products have historical antecedents framed around moralistic representations of the commodity and changes in the political economy of production.

Emerging 'public' spaces of consumption

Changes in consumption have nevertheless produced change in social-spatial relations. Modernity and the emergence of mass production and consumption invoked new pseudo-public spaces, particularly in urban areas.[6] These new civic spaces and privately owned 'public' spaces provided places for individuals to engage in 'conspicuous' consumption: to purchase, use and display commodities and to actively participate in the creation of commercial and classed cultures. Music hall, theatres, restaurants, ballrooms and dance halls, museums, tourist sites, fairground attractions and shopping precincts were just some of the consumption spaces which were popular in the nineteenth and early twentieth centuries. These spaces also marked the ascendancy of leisure as a structuring realm of everyday life, with media, marketing and advertising industries promoting narratives that connected pleasure and the purchase of commodities to being in, and being seen in, these spaces.

The department store was one of these new spaces of consumption; they were established in most cities in the Western world by the early twentieth century (Nava, 1997). While these stores did not constitute an 'everyday' shopping experience for many consumers, they assumed a critical role in shaping consumer culture, linking shopping with pleasure (rather than utility) and through a dramatic 'staging' of consumption which reinforced the symbolic attributes of goods (Laermans, 1993).

Department stores offered middle class women, 'who until the middle of the nineteenth century were confined to the private sphere, the opportunity to escape the dullness of domestic life' (1993: 94). Stores were advertised as spaces of legitimate indulgence, sites in which shopping was easy but also pleasurable, leisurely, romanticized and sensuous (Rappaport, 2000). The freedoms portrayed for women were paradoxical as department stores simultaneously reinforced and reproduced domestic ideologies and broader discourses of gender.

While offering women possibilities for democratization of consumption, the purpose of these spaces was to provide a space for women to make purchases for the home, thus reinforcing notions of domesticity (Bowlby, 1985).

Yet as Blomley (1996) observes in his reading of Emile Zola's *Au bonheur des dames*, a historical novel of a Parisian department store, such stores were also masculine spaces: men shopped, the majority of workers were men, and the spatial logic of stores was aligned with capitalist and patriarchal discourses enshrining male sexuality and identity.[7] Nevertheless, the representation of the department store as a glorified bourgeoisie household was intended to contain the potentially radical message of women indulging in public pleasures (Rappaport, 2000: 42), a message which was challenged by the New York woman and the middle class shoplifter, as we see in Box 2.4.

**BOX 2.4 'CONSUMING' WOMEN OUT OF PLACE:
THE NEW YORK WOMAN AND THE MIDDLE CLASS SHOPLIFTER**

Women's experiences in public spaces were a 'quintessential constituent of modernity' (Nava, 1997: 58). In 1860s New York, new public spaces had emerged in the form of public parks, restaurants, hotels, theatres and museums. These were spaces of 'leisure and pleasure' of consumption and display created by a burgeoning merchant class and the elite as the reflection of status and civic commitment (Domosh, 2001). However, the presence of bourgeois women in these new consumer spaces created anxieties which Mona Domosh (2001) believes were manifest in the construction of the 'New York woman'. The New York woman of the 1860s and 1870s (who, according to Domosh, was of a similar type to the Parisienne woman of the 1880s) occupied these new urban consumption spaces yet paradoxically was in many ways 'out of place'.

The New York woman was attractive and stylish, garbed in fashionable attire and 'given' to the pursuit of her own pleasure and passions. She was only deceptively beautiful and superficially respectable, being 'fictitious within' (*The New York Times,* cited in Domosh, 2001: 584) and frivolous by nature. The New York woman's days were occupied in public display and the pursuit of idle pleasure. When at home, she was engaged in beautification and attention to her wardrobe, with no care for children, 'hearth' or husband. Such women were dangerous because they tapped into public anxieties which rested on the need to ascertain a woman's 'true' nature, status and respectability, something which could not easily be 'read' in the façade of the New York woman (Domosh, 2001). Morality was embodied by the proper, bourgeois, white lady, in the drawing room at home, not women who spent their time engaging in consuming passions in spaces previously the domain of men.

Domosh (2001: 588) argues that consumption, with its indulgent, playful and leisure time association, can be contrasted with the utilitarianism, diligence and hard work necessary for production in modern industrial society. In her view the two conflicting value systems are reconciled by aligning the world of production with men, and the world of consumption with women. The New York woman subverted this naturally moral order by overconsuming and by consuming without wisdom or self-control. She upset not only nineteenth century gender ideology, 'but also the precarious balance between production and consumption and between self-control and self-indulgence' (2001: 590). Domosh's geographical research on the New York woman thus gives a fascinating insight into how spaces and identities are created and coded morally. Her work also has relevance for contemporary consumption. She believes that the mistrust of consumption and self-indulgence, and the association of the feminine with the excesses of consumption, are naturalized in descriptions of postmodern consumer spaces as superficial sites of pleasure and spectacle (see later in this chapter).

The contradictory nature of consumption was also evident in the treatment of middle class shoplifters in Victorian department stores (Abelson, 2000). Examining cases of female shoplifters in the United States, Abelson notes that women of the middle classes were allowed to shoplift and escape prosecution, in contrast to women of the working classes. Often 'a blind eye' was turned to middle class 'shoplifting ladies' (particularly those recognized as good customers). If they were caught, store owners were usually reluctant to prosecute them. When a case did go to court, 'female weakness' or kleptomania (constructed as an illness) was often accepted as a defence for the infringement. Kleptomaniacs were considered 'hysterics who had little rational control over their desire to consume' (Roberts, 1998: 818). This ideology of feminine weakness was created by the media, retailers and the women themselves, and was at times reinforced through the courts.

The discrepancy between the treatment of these women and that of working class women (who may have been more likely to steal from need) arises because the term 'shoplifting ladies' was seen as a contradiction. Middle class women who stole challenged existing stereotypical notions of how class, gender and crime were linked (Abelson, 2000: 310). Acting against middle class clientele was problematic as these women were the backbone of the department store institution. The cornucopia of commodities, their seductive display and the increased freedom and anonymity of the shopper meant storekeepers saw themselves as complicit in creating an environment in which shoplifting might seem like a invitation. Public prosecution was potentially injurious to business, even though many magistrates were 'openly resentful of a system that forced them to participate in such visibly class based justice' (2000: 313).

Abelson's discussion of 'shoplifting ladies' demonstrates the tensions between production and consumption. The more favourable treatment of middle class shoplifting women allowed the proponents of consumer capitalism to mask its contradictions. In many ways 'shoplifting ladies' were as 'out of place' as the 'New York woman'. The transgressions of female middle class shoplifters were downplayed to maintain the ideological hegemony of the gendered (and classed) construction of consumption. The New York woman was, by contrast, publicly vilified in order to achieve similar ends.

Fordist production and consumption

One of the key ways in which these changes have been understood is via an extension of Marxian work, examining in detail the structures and mechanisms inherent in the development of industrial capitalism. Under modernity, capitalism changed from a liberal to a *laissez-faire* model in the eighteenth and nineteenth centuries, and to state managed organized capitalism under Fordism (Lash and Urry, 1987). A Fordist regime of accumulation is theorized as having dominated production–consumption relations in advanced capitalist nations from the end of the Second World War to the 1970s. Fordism comprised a (relatively) stable alliance between capital, labour and the state, ensuring increases in production were matched by consumer spending, often with encouragement or direct intervention by the state (Lee, 1993). Medical, educational, energy and transportation services were often managed collectively and facilitated by the state and the local state. In New Zealand, for example, during the years of the welfare state from the 1930s till 1984, successive governments promulgated 'home ownership' as an ideal, building state homes and making available loans to prospective purchasers, thereby encouraging consequent consumerism of domestic and material possessions.

Taylorist work practices based in scientific management and rationalization[8] and forms of mass production based on the assembly line also underpinned the development of Fordist accumulation. Technological, transportation and communication advances which accompanied Fordism facilitated the speed of capital circulation, enabling the broadening and deepening of production and distribution networks (see Chapter 5). Consumption of standardized goods in mass markets became widespread, with commodification progressively encompassing more aspects and spaces of everyday life.

Explanations of changes in consumption which are based in Fordist regimes of accumulation tend however to posit consumption as an effect of production. Fordism itself has been demonstrated to be a fractured and much debated concept which has been unevenly manifested across space. Fordist explanations also tend to obscure the centrality of cultures of consumption in societal and spatial change (Lee, 1993). Lee argues Fordism was also concerned with establishing a 'social consciousness' based upon mass commodity consumption, a social consciousness articulated through the language of advertising and 'manipulated' by the media (1993: 88). Aestheticization, branding and standardization facilitated the development of new commodity aesthetics, many of which drew on modernist notions of scientific progress and technological knowledge (such as 'functionalism' and 'streamform' design). Consequently while the development of consumer culture is associated with significant changes in production (the rise of capitalist society, Fordist organization) it should be seen as a critical rather than a consequential part of modernity as it has developed in the West in the late twentieth century (Slater, 1997).[9]

Alternative geographies

A danger in presenting a chronology of consumption is that it is easy to assume consumption has followed a singular and universal trajectory differentiated simply in terms of rate of change, geographic extent and time (Glennie, 1995). Rather, consumption is contingent on place based and complex social, political and economic interactions between people, things and processes. The discourses and practices which comprise consumption are not universal but spatially and temporally uneven. It was mentioned earlier how, in early capitalist Britain, grain continued to be offered according to principles of need. The hegemony of capitalist systems of consumption and production has also been contested by 'alternative' forms of consumption which pre-existed and developed alongside capitalist systems of exchange (such as in socialist and barter systems of exchange: see Box 1.5). Box 2.5 discusses one such 'alternative' which emerged in nineteenth century Europe.

BOX 2.5 CONTESTING THE HEGEMONY OF CONSUMERISM: CONSUMER SOCIETIES

Martin Purvis (1998), in his paper in the *Journal of Historical Geography*, demonstrates how the 'emergence of a consumer society' was not unequivocally accepted by the 'masses', but was challenged by the creation of alternative infrastructures of distribution and production. The formation of societies of consumers in Britain and Europe between 1850 and 1920 was reflective of a growing desire amongst working people not simply for a larger share in the consumption of national resources, but also for greater control over production and distribution and hence potentially over supply and their own employers (1998: 149).

Consumer societies as cooperative attempts to collectively empower consumers became a means of defence against the powers of distributors and producers to create and discipline mass markets (1998: 155). However, though cooperative principles existed as an alternative to competitive capitalism, these societies could not fully separate themselves from consumerism or capitalist systems of accumulation. The provision of social or financial resources to the benefit of specific groups or individuals, ideological diversity and local particularism mitigated against the ideological belief of 'consumption as a shared identity' (1998: 164). The power of cooperative societies as agents of social-economic transformation was also reduced as consumer societies gradually acquiesced to state requests for transparency in activities and finances (in return for legal protection).

Nevertheless, while consumer societies' resistance to the domination of capitalist mechanisms of production, consumption and exchange was limited, the achievements of these societies in penetrating local retail markets (upwards of 30 per cent in some cases) and the benefits provided to members through trading, educational welfare and social activities were significant (1998: 164).

Postmodernity and Niche Consumption

In the mid 1970s a series of transformations in political, economic and social life heralded a new era of consumption oriented to serving niche rather than mass markets. In this most recent phase, the symbolic nature of consumption is said to have taken precedence over the commodity form, with media and advertisers playing a critical role in the manifestation of signs and images (Jackson and Taylor, 1996). Individual rather than collective forms of consumption predominate and consumer choice and identity become related to consumption of commodities as part of a *post*modern epoch or condition, a period differentiated from, but understood in relation to, the modern period which preceded it. The postmodern condition represents an emerging human world of 'flexibility, diversity and differentiation' which is reproduced in both society and space (Cloke et al., 1991: 179).

The term 'postmodernism' has been used in three ways: first to describe an epoch, an actual state of affairs or condition in society; second to define an artistic or architectural style; and third to refer to a set of ideas and attitudes (or methods) which relate to the former two constructs (Ward, 1997). Table 2.1 outlines some of the major themes which writers have focused on when discussing differences between modern and postmodern periods.[10] Identifying characteristics of a postmodern epoch is exceedingly difficult because postmodernism as a sphere of social life also defies characterization and encapsulates many contradictory phenomena.

There is considerable controversy over whether the postmodern condition actually exists (Slater, 1997) and whether the changes identified in Table 2.1 are simply an extension of earlier processes (such as a relationship between consumption practices and identity formation, and discourses of pleasure and sociality: Glennie and Thrift, 1992; 1996) or are qualitatively different. Harvey (1989) argues that postmodernity represents (and should be read as) another phase in capitalism. Expanding on the work of Marx, he asserts that postmodern places and processes can be understood as the latest expression of monopoly capitalism, a view which has been criticized for failing to approach consumption as a substantive topic in its own right (Jackson and Thrift, 1995).

The identification of a postmodern epoch is said to rest on three main processes connected with consumption: commodification, social division and new forms of everyday life (Glennie and Thrift, 1992). Drawing on some of the themes identified in Table 2.1, these processes will be explored in turn.

Commodification and commodity fetishism are presumed to have penetrated virtually every aspect of the geographies of everyday life in a postmodern epoch. It is argued that signs and images take on the logic of the commodity as they start to become commodities in themselves. Jean Baudrillard (1983) believes that commodities get their meaning from their position in networks of commodity signs. In other words, 'things' can only be understood in relation

TABLE 2.1 *Characteristics of the postmodern epoch*

Modernity (*c.* 1750–1970s)	Postmodernity (*c.* 1970 onwards)
Technological superiority and progress celebrated	Technological superiority and progress challenged
Sameness, universalism	Difference, diversity, discontinuity, fragmentation
Search for absolute knowledge	Belief in relative knowledges
Depth and essence	Surface, hyper-reality
Rules and regulation of style	Pastiche, collage and spectacle
Belief in real authentic world that exists outside our knowledge of it	Idea of independent authentic reality challenged; the relationship between the 'real' and how this is represented is not simple or straightforward
Mass production of commodities (i.e. in large batches and runs) for consumption by mass markets	Niche production of commodities occurring in short runs and small batches specifically for consumption in differentiated markets
Production of objects economically and politically significant	Consumption and reproduction of images economically and politically significant (rise of social movements connected to patterns of consumption)
State power, welfarism and intervention	Decline of state power, individualism and enterprise
Goal of individuals is concerned with satisfying wants through material possessions	Goal of individuals is attaining desires through symbolic meanings and images of possessions
Emphasis on character	Emphasis on self and bodily representation
Identity formed in relation to sphere of production (e.g., work) and collectively based identities have a degree of stability	Identity formed in relation to sphere of consumption (e.g. leisure) and individualized, fragmented, multiple, shifting and contradictory identities of subject

Sources: drawn from Cloke, 1993; Cloke et al., 1991; Featherstone, 1991; Glennie and Thrift, 1992; Harvey, 1989; Lash and Urry, 1987; Rojek, 1995; Shurmer-Smith and Hannam, 1994; Ward, 1997.

to other commodities. The invasion of signs and images and the multiplication of cultural mediators in everyday life (such as electronically mediated signs, images, spectacles and simulations appearing on television and via the Internet, on billboards, at sports grounds and so on) are seen as symptomatic of stronger connections between commodities and identity value than previously existed. Commodities can be used by individual subjects as a signal of both individuality and commonality with others because identity is related to how and what people purchase, use and display and the commodity practices associated with this (though Lodziak, 2000, claims this connection has been exaggerated).

In a postmodern epoch, consumption has come to be regarded as a symbolic rather than an instrumental activity (Campbell, 1995: 99). Debord (1994), for example, suggests life is experienced as a series of spectacles, while Baudrillard (1983) argues a new regime of simulation characterizes the postmodern epoch. Fantasy, spectacle and simulation may be highly visible in built environments such as theme parks or urban and harbour redevelopments, but are also manifest in the mundane activities of everyday life (watching television, being 'entertained' while standing in a bank queue).

The intensification of divisions between individuals and groups is assumed to be another major consequence of changes in consumption in a postmodern epoch. Producers create commodities which are marketed to serve (and also create) diverse sets of consumers and consumption practices. In order to maintain social divisions (and ensure continued growth of markets), fashion, advertising and design help to stabilize the meaning of goods, facilitating the reproduction of appropriate structures of taste and practices of consuming (Glennie and Thrift, 1992). In a postmodern epoch 'communities of consumption replace real community, public life gives way to organized commercial spectacles', anonymous transactions predominate, and life is experienced as challenges to be overcome by the acquisition of commodities (Bauman, 1990: 204, cited in Goss, 1999a: 114). The popularity of retro music (for example, the comeback of the music of Abba and the Bee Gees), and the desire for authentic experiences and places (as in forms of tourism: see Figure 2.3), are viewed as attempts to escape this simulated world by making a fetish of the supposedly real world (one which Baudrillard argues does not actually exist). Thus, in a postmodern epoch, consumption experiences are said to take on a form of hyper-reality, where the representations assume a reality of their own. Retaining 'a sense of place' can be viewed as a means of coping with instability and uncertainty of existence in a postmodern condition.

Associated with this is the establishment of a post-Fordist regime of accumulation centred on flexibility in products and production systems. Meeting the desires of niche consumers involves the production of smaller batches of commodities, tailored to suit specific cultural groups, lifestyles, labour markets and places.

In a postmodern epoch, social division is assumed to be less associated with the sphere of production (a particular job of 'housewife' or 'farmer', or a collective grouping such as middle or working class) and more closely attached to the sphere of consumption and the realms of leisure, pleasure, fashion and desire. The stability of identity is gone[11] as a postmodern society is also a 'risk society' no longer offering the securities, stabilities, associations and essentialisms which once existed (Beck, 1992). There are many identities to choose from and it may be impossible to separate one's construction of self from the identities portrayed in film, advertising, fashion, magazines or fitness programmes

FIGURE 2.3   The search for authentic tourist experiences (here an elephant ride in the jungles of Thailand) is seen as part of a search for the authentic in a postmodern epoch

(Ward, 1997). Consumption becomes a means both of self-reflexive distinction and self-actualization and of belonging (Shields, 1992a), with commodities an important means of establishing different body images, lifestyles or social types (see Box 2.6). Featherstone (1991) suggests that post-Fordism has created a new middle class who favour narcissistic, self-conscious and stylized consumption practices.

BOX 2.6 POSTMODERNITY AND LIFESTYLE SHOPPING

Rob Shields' edited collection on *Lifestyle Shopping* (1992b) sought to investigate how new modes of subjectivity were emerging as part of a theorized postmodern condition. Shields' emphasis was not so much on the representational aspects of places but on the role of consumer culture in 'the private spaces of subjectivity, media and commodity consumption, and the changing spatial contests of everyday public life' (1992b: 1). Lifestyle shopping in carnivelesque sites was seen as providing a means where new subjectivities could be tried out, displayed and discarded. Shields viewed the combination of leisure and consumption activities in particular sites as a shift from modernity. He saw shopping as offering multiple personas and identifications to consumers as members of consumer tribes, centred upon lifestyles comprising behaviour, adornment, taste and habitus (drawing on Bourdieu: see Box 4.1). Consumption thus becomes not 'merely the means to a lifestyle, but the enactment of lifestyle' (1992b: 16). While notions of 'lifestyle shopping' have been debated by geographers, Shields' discussion of resistance has been frequently cited. Shields noted how the existence of consumption sites as spaces of leisure, carnival and festival existed in tension with the commercial aspects. These tensions were not easily overcome, with the need for managers of these spaces to limit and/or control, the carnivalesque, the difficulties posed by people who shop but do not purchase, and the possibilities of consumer resistance.

A third change, Glennie and Thrift (1992) assert, has been the development of new forms of everyday life which are expressed in ways of being, visual landscapes and built environments. The exponential growth of the Internet and the circulation of information by electronic means have been significant in terms of communication and consumption (see Chapter 3). Everyday life is assumed to have become more reflexive and aesthetic, with human subjects possessing an increased ability to reflect upon the social conditions of their own existence (1992: 436). Glennie and Thrift (1996) (whose work was examined in Box 2.1) challenge the 'postmodern thesis' where commodities are seen as increasingly disposable and meaningless,[12] arguing greater reflexivity does not necessarily imply greater anxiety and alienation or loss of authenticity, but instead may invoke possibilities for the reconstitution of human meaning. They believe cultural segmentation of consumers rather than fragmentation is occurring, arguing that the number, scale and power of social groupings committed to particular consumption patterns has grown (think, for example, of the proliferation of food and diet related consumer movements).

'Placing' Transformation in Consumption: Problems of Extrapolation and Interpretation

There is considerable debate over the viability of the three-stage model (the emergence of modern consumption; consolidation and mass consumption; and postmodern consumption) identified in this chapter. When 'historical perspectives on consumption are used to conceptualize contemporary consumption' they may 'inaccurately categorize consumption past and present' (Glennie and Thrift, 1992: 423), thereby failing to clarify which facets of contemporary consumption represent 'continuities with earlier practices and those which are genuinely novel' (Thrift and Glennie, 1993: 33). The real challenge, according to Glennie (1995), is to think about what is distinctive about consumption in different periods. A further difficulty with the chronology presented here is the presentation of modernity as a defining feature of modern and postmodern consumption, and consumerism itself. Trentmann (2004: 374) argues this result in a view of change which fails to 'problematize how (and whether) these different types have emerged, developed, and stood in relation to each other in different societies at different times'. When consumerism is conceptualized as a type of mentality or behaviour directed towards the acquisition of material goods, then it is no surprise that it becomes a fundamental feature of western modernity (2004). In Trentmann's view this construction of consumerism denies other forms of consumption and sociality (such as practices based around services and experiences), and ignores the significance of non-commercialized consumption activities.

In addition, the three-stage model outlined in this chapter has been formulated primarily around the experiences of Western countries and consumers. The experiences of countries in the Third World, or in the so-called newly industrializing countries (NICs) of Asia, for example, cannot be assumed to have altered along similar trajectories. Box 2.7 explores the Russian Federation (previously part of the USSR) to examine how consuming has been rapidly reconstituted with the shift from a centrally planned socialist economy to a state managed capitalist economy over the last decade.[13]

BOX 2.7 CAPITALISM, COMMODIFICATION AND CONSUMPTION IN RUSSIA

From 1989 and the fall of the Berlin Wall, a number of previously socialist countries in Eastern Europe sought to open their economies directly to the influence and operation of the capitalist system. The opening of Russia's economy to global capitalism in the 1980s and 1990s has not mirrored the experiences of newly emerging capitalist economies in the eighteenth and nineteenth centuries. Kapucsinski states: 'The advance guard of the capitalism that arrived in Moscow consisted of armies of speculators, barons of the black market, gangs of drug dealers, armed aggressive racketeers, brutal, ruthless, powerful mafias' (1996: 55). Social and economic problems were accentuated as the old system of state organized production and consumption was dismantled without new institutions and structures being put into place to ease the adjustment from a socialist to a market economy (Nelan, 1991). Prices rose 48 per cent by June 1991, consumer goods became scarcer and lines for basic items in cities lengthened (1991: 18). In 1993 conditions began to improve and a period of slow but shaky stabilization was under way. The country's currency crisis in 1998 and a chronically inefficient distribution infrastructure severely limited the range, quantity and quality of consumer goods available. Under the current presidency of Vladamir Putin, inflation has continue to remain high at around 17 per cent (*Moscow Times,* 2001: 6) but economic growth has improved (8 per cent in 2001: Binyon, 2002a: 8). Producer services have grown significantly and many companies in Russia are now foreign owned; Western owned multinationals dominate advertising agencies, for example (Schofield, 2001: 7)

The rise of mafias and official corruption has undermined the morale of citizens (Argenbright, 1999: 2), with high levels of crime, money laundering, and drug and alcohol abuse remaining intractable problems (difficulties which are not unique to Russia or to countries with a post-communist legacy). Despite achievements in industrial production, in 1995 approximately 26 per cent of the population had gross incomes below the official poverty line, compared with 2–5 per cent before the reforms (Tikhomirov, 1996: 1).

At present a strange hybrid of state, market and shadow or informal economy forms coexists. Mellow (1997) argues that a new capitalist class has emerged, a group of predominantly young financiers, their income and influence providing access to a huge range of consumer goods and services from within and outside Russia. The new Russian middle class is buying more Western products than ever, including $10 billion of imported foodstuffs alone in 1996 (1997: 54).

In Moscow,[14] shortages of commodities and queues are no longer as prevalent as in the early 1990s. However, access to consumer goods and new consumer spaces in many rural and remote areas has not altered substantially, with the purchase of basic commodities still remaining difficult for many Russians. Thousands of shoppers from the 'burgeoning middle classes' might be 'streaming into the glittering new shopping centers where formerly only foreigners and wives of oligarchs were seen' (Binyon, 2002a: 8) but an estimated 300,000 people are excluded from such spaces, living homeless in Moscow, with 50,000 of these being children (Binyon, 2002b: 28). A survey of Muscovites (Bater et al., 1995: 686) revealed the reform process had made life more difficult in a material sense, if not an emotional one, and there was a deep frustration with daily life, with the freedom to acquire consumer goods becoming a source of great divisiveness (Kapucsinski, 1996).

Dramatic and rapid transformation has also occurred in the landscape of major urban areas as the state has retreated from the public space of the street. Argenbright (1999) suggests the open and welcoming design of the first McDonald's (opened in Moscow in 1990) was duplicated by numerous consumer outlets. Since the break-up of the Soviet Union, Western commercial cultures are increasingly visible; fast food restaurants and kiosks, sidewalk cafés, movie theatres and new hotels have become a ubiquitous part of the city's landscape (Baker, 2001: C16). In Moscow, advertising is commonplace (see Figure 2.4) and footpaths, subway entrances and pedestrian tunnels have been colonized by retailers and hawkers (Argenbright, 1999).

FIGURE 2.4 Advertising in post-Soviet Russia is now a significant feature of the Moscow urban landscape

> Visible transformation in post-Soviet Russia does not in itself provide an understanding of how consumption is constituted or experienced by Russian consumers. Open-air markets which existed in communist times continue to play a role in compensating for the dysfunction of redistributive systems, filling an important role as a supplement to and substitute for the formal economy (Sik and Wallace, 1999). New consumer spaces and practices have also had a 'productive' role in the creation of new forms of commercial culture, such as in the growth of the nightlife dance/rave and youth subculture (Wallace and Kovacheva, 1996) and 'pet culture' in which animals become markers of status and identity (Barker, 1999). The experiences and 'imaginings' of Soviet consumption also continue to influence meanings of contemporary consumption practices and expectations (Oushakine, 2000). Oushakine's research with high school and university students in Siberia revealed, for example, that though the old Soviet lifestyle was viewed by participants as restrictive and limiting it was seen as encompassing a wider variety of consumption practices (theatre, museums, newspaper reading and so on), whereas the new Russian style of consumption, though more relaxed and enjoyable, was reduced exclusively to shopping (2000: 101). These experiences were constituted in relation to consumerism in pre- and post-Soviet Russia rather than duplicating those associated with Western consumerism.
>
> Thus social transformation in production and consumption never takes place on a 'blank slate' (Kollantai, 1999). The case of Russia illustrates well the hybridity of forms, practices and constructions of consumption that exist, the disjunctures that can occur, and the way in which consumption is actively made and imagined rather than reproduced to a standard historical framework or trajectory in place.

Histories and Chronologies: the Need for Specificity

Historical change in consumption has been discussed as a chronology, a term which implies a specific and agreed trajectory with points of origin and completion. The broad features of the three-part chronology presented in this chapter involve the emergence of 'modern' consumption in the seventeenth and eighteenth centuries, its consolidation and intensification in the nineteenth and twentieth centuries, and its increased prominence in the maintenance of daily life as part of a postmodern condition which emerged in the late twentieth century. In this schema, consumption changes are frequently associated with major economic transformations during these periods (industrialization and urbanization, the geographical extensions of capitalist relations of exchange, Fordist mass production and the welfare state, post-Fordist flexible accumulation). However, it is fallacious to assume that transformations in consumption are simply a consequence of changes in the sphere of production. In a postmodern epoch there is also a tendency to see consumption and consumer culture as separate and fragmented narratives with no coherence outside themselves, leading to the unhelpful detachment of consumption behaviour from everyday life and 'attributing too much autonomy to culture from economic, political and social contexts' (Glennie, 1995: 191).

Generalized models are useful for mapping contours of change, and the threefold chronology presented here should be used as a basis to ask questions about consumption places and processes. There is a need to interrogate such chronologies to explore their hegemony and their relevance across different commodities, consumption systems and commercial cultures, discourses of consumption, and consumer practices and experiences in various spatial and temporal contexts. It is vital to examine whether (and how) they are articulated in particular time and place contexts at a variety of scales (for example, home, rural, urban, community, nation-state), to consider how such changes are invoked, made meaningful, embodied and constituted in everyday life by differing groups and individuals, and to explore the ways in which these actions might be structured, constrained and enabled.

Satisfactory analyses of consumption practices should recognize that features of place, context and setting are central to many types of consumption and shape consumers' practices (Glennie and Thrift, 1993: 605). Although a considerable volume of material theorizing and analysing historical change in consumption has been written by historians, anthropologists, sociologists and economists, there appear to be fewer contributions by geographers. The relative lack of geographical work on historical transformation is perhaps not surprising given the huge volume of geographical research which examines consumption in specific social-spatial contexts and processes (primarily exploring contemporary settings with some historical work). Geographers have highlighted the difference place makes to how consumption practices are experienced and reproduced and how consumption processes and practices in turn create particular geographies. These ideas are examined in the next chapter.

FURTHER READINGS

Blomley, N. (1996) '"I'd like to dress her all over": masculinity, power and retail space', in N. Wrigley and M. Lowe (eds), *Retailing Consumption and Capital*. Harlow: Longman. pp. 235–56.

Domosh, M. (2001) 'The "Women of New York": a fashionable moral geography', *Environment and Planning D: Society and Space*, 19: 573–92.

Glennie, P.D. and Thrift, N.J. (1992) 'Modernity, urbanism, and modern consumption', *Environment and Planning D: Society and Space*, 10 (4): 423–43.

McKendrick, N., Brewer, J. and Plumb, J.H. (1982) *The Birth of a Consumer Society: the Commercialization of Eighteenth-Century England*. Bloomington, IN: Indiana University Press.

Purvis, M. (2003) 'Societies of consumers and consumer societies', in D.B. Clarke, M.A. Doel and K.M.L. Housiaux (eds), *The Consumption Reader*. London: Routledge. pp. 69–76.

Shields, R. (ed.) (1992b) *Lifestyle Shopping: the Subject of Consumption*. London: Routledge.

Trentmann, F. (2004) 'Beyond consumerism: new historical perspectives on consumption', *Journal of Contemporary History*, 39 (3): 373–401.

NOTES

1 Campbell (1987:201) also argues that Romanticism provides a context for discussing contemporary cultural change. In contemporary society, he suggests, the emphasis on self-expression and self-realization is fundamentally 'Romantic' in inspiration.
2 Miller (2001c) suggests negative moralities continue to underpin contemporary consumption practices, defining both 'morally' appropriate forms of consuming and ethical consumption practices and relations toward others (see Chapter 7).
3 The ideas of Walter Benjamin and Pierre Bourdieu, which will be outlined in later chapters, also provide further insight into aspects of the cultural production of consumption and the creation of commercial cultures.
4 Veblen has also been criticized for his positioning of women as subordinate to men and as superfluous and costly consumers (Edwards, 2000).
5 The work–leisure, production–consumption dichotomy is one which masks the production inherent in consumption and vice versa, and one which in a problematic way locates (non-productive) pleasurable consumption in specific places (for example, the domestic sphere, home, the sports field) and by implication defines the consumption premise of certain groups (the unpaid, women etc.). It is also problematic in terms of the conception of leisure as residual or spare time left over from work, and again is suggestive of consumption as always associated with leisure or the frivolous pursuit of pleasure (made legitimate through paid work) rather than as an active, productive and powerful means of engaging in everyday life.
6 The word 'pseudo-public' is used because many of these public spaces were actually privately owned and commodified.
7 Blomley's (1996) fascinating exposition of Zola's novel also exposes the way in which economic and cultural logics of consumption are inseparable.
8 The term 'Taylorism' derives from Frederick W. Taylor's *Principles of Scientific Management* (1967 [1911]). This involves breaking down a task into its simplest components and organizing these fragmented components along 'rigorous standards of time and motion study' aimed at achieving efficiency and productivity (Harvey, 1989: 125). Fordism as a method of industrial production incorporated such methods with new technologies of production and mechanization.
9 Modernity had its origins in seventeenth century Europe and was based on the so-called 'Enlightenment ideals' of faith in progress, optimism, rationality, the search for absolute knowledge and the knowledge of true self which was the foundation for all other knowledge (Ward, 1997). By the mid twentieth century it is hypothesized that modernity became the dominant social order globally.

10 Compiling such a table is in itself 'risky' because the simple act of labelling a set of changes as 'postmodern' suggests that there is an essential (and agreed) truth about what the postmodern epoch comprises and that these changes can be somehow separated from the postmodern attitudes and ideas that inform them. Some postmodernist thinkers might see the oppositional categorization represented in Table 2.1 as based on modernist ideas about the rational construction of knowledge. Thus the form and nature of such changes may be viewed very differently from those influenced by postmodern attitudes and those influenced by other philosophical traditions. In labelling a characteristic 'modern' or 'postmodern' the table also tends to suggest that the particular condition or state of affairs has shifted from one stage to another and is experienced everywhere and in the same way.

11 Some postmodern authors believe 'real' unified selves never existed (Ward, 1997).

12 Indeed much of the emphasis of this book is on how consumption is a meaning 'full' rather than a meaningless sphere of actions, practices, relationships and spaces.

13 It is not my intention, nor would it be possible within the confines of a brief case study, to document, account for and analyse the changes in consumption in the post-Soviet era.

14 In many regards Moscow, as the chief city at the apex of the urban political system, is unrepresentative of other Russian cities (Bater et al., 1995). In 1996 two-thirds of all foreign investment went to Moscow (compared with St Petersburg at 4 per cent for the same period) and 80 per cent of finance capital is reportedly based in the city (Argenbright, 1999: 4).

3

Spaces

As indicated in Chapter 1, geography matters, not just because consumption takes place in space, but because space is produced through consumption. Geographers have focused on place, scale, context and spatial organization of consumption to understand how places and processes are mutually constituted. In doing so they have examined how places in which consumption occurs are social spaces, influencing the formation of social relations and subjectivities. This chapter explores the interrelationships between social, cultural, political and economic processes as a way of understanding how consumption spaces, scale and spatialities are created and the geometries of power associated with them.

Space, Place and Scale

In recent years the way in which geographers have used concepts such as 'place', 'space' and 'scale' in human geography has been debated.[1] Geographers such as Massey (1984), Harvey (1982) and Soja (1989) have long argued that space should not be viewed as a universal, absolute and neutral container in which objects are 'placed' and events occur, nor should social relations be treated and mapped simply as spatial relations. Instead space should be viewed as socially produced, as given meaning through human endeavour (see Box 3.1). If spaces are socially produced, then according to Massey (1993) places are not simply areas on maps but shifting bundles of social-spatial relations which are maintained by the exercise of power relations.

BOX 3.1 LEFEBVRE AND SOJA: MAKING SENSE OF SPACE

Henri Lefebvre's (1991) characterization of spatiality explains how space is socially produced as both symbolic and material space. Lefebvre's ideas centred upon the capitalist production of space. He argued that globalization and modernity result in the colonization of the 'concrete' space of everyday life by abstract space (visualized space which has no social existence) (Gregory, 2000). Space in Lefebvre's schema is not a blank slate but is actively involved in the production of social relations and new forms of spatialities.

Geographer Edward Soja (1996) expanded Lefebvre's threefold conceptualization of social-spatial relations. Soja's 'trialectic' consists of interconnected realms: spatial practices, representations of space, and spaces of representation. *Spatial practices* are social processes which reproduce spatiality and which are both the medium and the outcome of human activity (ranging from individual practices such as shopping to collective and institutional practices such as consuming health services and the state regulation of retail). *Representations of space* comprise the space of the imagination which is the conceived and ideological space of people encapsulated in the visions of designers, planners, artists and architects (for example, in maps, designs, paintings). Representations of space can become powerful and tend to become real geographies as spatial practices (such as commodification).

The third member of the trialectic is *spaces of representation.* These are the real and imagined geographies as they are lived by inhabitants and users, who must translate and negotiate representations of space (which may 'colonize' the lived world). Soja (1996) envisaged possibilities for resistance, emancipation and change in viewing representations of space as part of a third space, a simultaneously real and imagined space in which oppositional categories which define social–spatial relations (and space itself), such as racism, patriarchy and colonialism, might be overcome and disrupted.

Geographers concerned with reading landscape have sought to analyse and explore the significance of representations embedded in consumption spaces for both users and producers of these spaces. Their work has provided insights into the contexts of commodities (Sack, 1992) and how geographies of consumption are themselves consumed by people who frequent these spaces (see Boxes 3.2 and 3.5).

Scale differentiates space. Scale is not space *per se*, nor is it a static hierarchy placed over existing space; it should be seen as mobile and changing, a 'relational element in a complex mix that also includes space, place and environment' which embodies relations of empowerment and disempowerment (Marston, 2000: 221). Consequently the production of scale is an integral part of the production of space, and spatiality is a function of the difference between places. Such differences may take the form of emotional, discursive, material, technological or instructional dimensions of landscapes (Amin, 2002).

Discussions of the production of place and scale have tended in Marston's (2000) view to be framed around discourses of production which centre on the structures and actions of state, capital, labour and non-state political agents at local, regional, national and global scales. Yet consumption is integral to the production of scale as a 'powerful and pervasive place building process' which is central to the everyday experiences of those in the Western world (Sack, 1988: 643; and see Box 3.2).

BOX 3.2 ROBERT SACK AND THE CONSUMER'S WORLD

The publication of Robert Sack's *Place, Modernity, and the Consumer's World* in 1992 highlighted consumption as a place forming and place altering act. Sack developed what he called 'a relational framework' which used a loom as a metaphor for the formation of the consumer's world. Consumers weave together threads (elements from nature, meaning and social relations which are contained within commodities) and in consuming commodities create 'the fabric' which is the context of consumption. Advertising in Sack's view has a critical role in creating the context of the commodity, operating as a language which allows consumers to weave idealized worlds which 'reveal how products are supposed to affect our lives' whereby 'these messages become part of people's attitudes towards the actual products that appear in real places' (1988: 643).

Sack's consumer world primarily consists of landscapes produced by or for mass consumption, that is, department stores, retail chain stores, shopping malls, the home, tourism landscapes, workplaces, theme parks and museums. These are presented as disorienting places which function as settings for the display and purchase of commodities embodying tensions between reality and fantasy, authenticity and inauthenticity. Commodities are seen as joining private (subjective and idiosyncratic space) with public (objective and alienating) space (1992: 134), transforming places into stages or settings in which people can express themselves, while attempting to overcome the alienation and disorientation characteristic of the postmodern condition. The more commodities consumed, 'the more selves we express' (1992: 153). Yet the expression of self(ves) in place is limited; the meaning of commodities is often hidden from consumers in consumer places by the 'front stage of consumption', with extraction, production processes, distribution, waste and pollution remaining 'backstage' (1992: 104).

Sack argues advertising creates meanings of what everyday places should be like; advertising is 'a blueprint of how commodities create places', with places of consumption being 'three-dimensional advertisements that not only sell commodities, but sell themselves' (1992: 133). However, Sack has been criticized for overemphasizing the role of advertising as the language of consumption and for his assumption that a view from nowhere (that is abstract-objective space) can actually exist (Warf, 1994).

Despite the criticisms levelled at Sack's (1992) treatise on consumption, its publication clearly signalled the importance of spatial practices and representations of consuming in everyday life and landscape at a time when geography literature was dominated by 'production' oriented accounts of space and place. He also established some important links to the consuming and the construction of place as a moral act, and to the moralities of geographical perspectives (see Chapter 7). Sack condemned what he saw as the obfuscation of the effects of consumption, which undermined humans' ability to act morally and responsibly, a theme which continues to underpin his work on consumption.

Spectacular spaces

Much of the initial work on spaces of consumption has focused on visible and spectacular spaces: hotel complexes, shopping malls, theme parks, heritage sites, casinos, sites of carnival and festival. The characteristics of such spaces – enclosed and separated from wider social spheres, centred around leisure, consuming and simulation, regulated by disciplinary technologies of surveillance, gatekeeping and crowds – are said to be disorienting, fragmenting and anxiety producing (Woodward et al., 2000). As spaces of representation these landscapes are argued to obscure the relations of capitalist production behind the aura of the commodities (Goss, 1999a). Ritzer (1999) suggests that such spaces are 'enchanted spaces' – spaces where inauthentic alienating experiences are replaced with magical manufactured leisure and consumption experiences, in order to make increasingly rationalized spaces (and the systems through which they operate) seem appealing.[2] Though such spaces are seen as symptomatic of the postmodern condition, they must be understood as contextualized and meaning laden landscapes, something Warren highlights in her discussion of Disneyland Paris (see Box 3.3).

BOX 3.3 THEME PARKS: SPECTACULAR SPACES OF CONSUMPTION

Theme parks are an example of the spectacular leisure and consumption spaces that have attracted the attention of consumption researchers (Sorkin, 1992). Stacey Warren (1999) endeavours to draw on narratives produced by visitors and workers to suggest that even these 'spectacular' spaces should not be viewed as sites of homogeneous experience or meaning. In studying cultural contestation at Disneyland Paris, Warren notes how the cultural and ethic layerings of meaning and practice have meant Disneyland Paris – the most popular tourist attraction in France – has 'become an intriguing hybrid of conflicting French, European, and American values' (1999: 110).

Disney's recolonization of space outside Paris was designed to encourage visitors to shed associations with the real world when they enter parks (1999: 114), creating a placeless environment which designs out contestable possibilities of non-Western discourses, and repackaging the colonial fantasies of the American company Disney for European consumers. While Disney produced a European form of its American 'magic' in the built environment of the park, Warren suggests it remains 'Disney' in operation, intellect and taste (1999: 117). Despite this 'the reality of the French theme park often strayed from the intended image' (1999: 118), undermining the stability of Disney's colonial project. The granting of concessions by the company on employee training, behaviour and dress code and on linguistic interpretation demonstrates how local consumers and workers in such sites can 'successfully challenge the supremacy of corporate culture' (1999: 123). Warren's narrative of Disneyland Paris demonstrates how cultural conflicts since the park's inception echo wider 'post-colonial struggle in the so-called real world' (1999: 112).

Trajectories of Consumption? Placing Consumption Sites

Geographers have also begun to examine a range of other and more mundane types of consumption space, and the insights gained from these spaces remain important to understandings of consumption as representational and lived spaces. Sites of consumption have tended to be discussed in terms of a historical trajectory, with arcades, department stores and shopping malls materializing consecutively as 'primary' sites of consumption over the last two centuries (Box 3.4).

BOX 3.4 BENJAMIN AND THE ARCADES PROJECT

Walter Benjamin's writings (1970; 1983; see especially Buck-Morss, 1989) explored nineteenth century European arcade culture. Benjamin saw these aesthetic covered spaces as 'dream worlds' in which middle class women in particular could shop and engage with the symbolic aspects of consumer goods and spaces (Edwards, 2000: 21). Benjamin argued the world of mass produced commodities was about illusion, and that value was eclipsed in representation (Nava, 1997). The reproduction of commodities violated individuals' sense of authenticity and reality (Benjamin, 1970), something the post-Enlightenment 'modern' world needed to recognize in order to connect with its unconscious (dreaming) potential (Edwards, 2000). Central to Benjamin's ethnography of arcades was the concept of the flâneur,[3] whose purpose in strolling public spaces was to see and to enjoy being seen. In the late nineteenth and early twentieth centuries women were encouraged to experience public (city) life in the role of the masculine character of the flâneur (Rappaport, 2000: 39), looking at and being with strangers. Benjamin's ideas have been reflected in writings on the department store and mall which focus on these spaces as sites of representation, as sites of individual and collective dreaming and as spectacles to be gazed upon and consumed.

The formation of department stores in the later nineteenth and early twentieth centuries was discussed in Chapter 2. Like Benjamin's arcades, department stores were sites of individual and collective dreaming (for a predominantly middle class patronage). However, there is a contradiction between the construction of the shopper as a masculine flâneur strolling and gazing in these spaces, and shopping as a social, skilled and knowledge based activity (Glennie and Thrift, 1996: 225). This contradiction is particularly evident in the construction of department stores as feminine spaces (see Figure 3.1). As noted in Chapter 2, department stores taught women to be consumers and provided spaces for 'everyday' encounters with commodities (Bowlby, 1985). Department stores were places where women could feel safe, empowered and free (to be and to dream) and would occupy roles other than that demanded by their

FIGURE 3.1　The interior of a department store in 1920s New Zealand. Such stores were seen as providing acceptable public spaces for women to shop and socialize (permission by Alexander Turnbull Library, Wellington, New Zealand from the Gordon Burt Collection)

family (Fiske, 1989: 20), yet discourses of family (naturalizing nuclear family and women as mothers and wives) and modernity (where scientific rationality, technology and functionality are valued) continued to position women as purchasers of household commodities (Dowling, 1993). Thus in addition to understanding how stores operate as representational spaces, research on the department store has provided insights into how discourses and subjects of consumption become attached to particular places and times, the effects of which extend beyond their physical boundaries (Dowling, 1993; Winship, 2000).

Shopping malls: all-consuming spaces?

As visible and often spectacular built environments, it is not surprising shopping centres have attracted the interest of consumption geographers. Shopping centres or malls[4] consist of a range of retail outlets and entertainment facilities within an enclosed space that is usually privately owned and managed. Since the first architecturally designed shopping mall was built in the United States in 1956, shopping malls have become a ubiquitous feature of the landscapes of urban areas in both developed and developing countries (Hedman and Sidel, 2000).

Many of the early shopping malls were stand-alone modernist 'machines for shopping' built on out-of-town sites competing directly with high street retailers and extending the one-stop-shop format of the department store (Bowler, 1995: 16). In the last two decades many more malls have been created in city centres and are re-created in a variety of nostalgic and contemporary architectural forms.

Initial work on shopping malls by geographers sought to investigate how these spaces operated as representational spaces of consumerism, based on a critical semiotic reading of the mall landscape. Malls have been understood as theatres of consumption with contrived sets designed to promote a 'retail drama' (Hopkins, 1991: 270), as leisure spaces and as tourist attractions (Butler, 1991), as sites of collective dreaming, pleasure and diversion (Backes, 1997) and as sites of 'strolling' for the flâneur. The representational space of the mall is symptomatic of commodification and part of the growing intrusion of spectacle, fantasy and escapism into urban landscapes (Hopkins, 1990). Malls are implicated in the production of nostalgia (for past places and times and the 'authentic'), carnivalesque (drawing on Bakhtin, 1984), magic and fantasy. Their built environments purvey myths of 'elsewhereness' (Hopkins, 1990): the overt manipulation of time and/or space to evoke idealized experiences of other places which create an illusion of a world standing 'outside of everyday life' (Goss, 1999b: 45). John Goss' work was instrumenital in highlighting the 'taken-for-grantedness' and power of retail landscapes (see Box 3.5).

BOX 3.5 JOHN GOSS AND THE MAGIC OF THE MALL

In a paper on the 'magic of the mall' Goss (1993) examined the form, function and meaning of shopping malls to understand how developers and designers of malls encourage consumption (or more specifically purchase). Drawing on North American shopping experiences, Goss' paper sought to argue that producers of malls designed the built environment of shopping centres to 'assuage collective guilt over conspicuous consumption' (1993: 19). This was done by mystifying (or obfuscating) the link between shopping and purchase in place. Commodities are severed from their relations of production and presented in terms of their potential transformative (and symbolic) value. Obscuring both relations of production and the capitalist logic of accumulation through commodities, exchange value might be realized. In order to achieve this, pleasure, fantasy and magic are promoted in shopping centre landscape through architecture, interior design and theming which draws on other times and places. The staging of events (fashion shows, carnivalesque entertainment) and services which facilitate consumption (such as food courts, day care facilities) also help in creating a liminal space, where the mundane is suspended and freedom to express oneself (by consuming commodities) is assumed free of censure. Goss highlighted the ways in which freedoms to consume are symbolic, with a politics of exclusion operating which means

> consumers are subject to rigorous surveillance. Though his work has been subject to criticism (see Box 3.6), his research on mall landscapes (Goss, 1999a; 1999b) has been instrumental in highlighting the power laden nature of consumption and the way in which places (both imagined and real, near and far) are actively constituted by the practices of mall owners, designers and retailers.

Attention has also focused on specific managerial and retail techniques and practices whereby consumers' desires and subjectivities are 'shaped up' in retail spaces (Hopkins, 1990: 15; Winchester, 1992). Sights, smells, sounds and music, for example, are important in evoking pleasurable consuming associations in shopping spaces (De Nora and Belcher, 2000). Bowler's (1995) interviews with Australian and New Zealand shopping centre managers demonstrated how spatial strategies of professionals are inseparable from understandings of malls as personally and socially constructed places, from the professionals' roles as consumers themselves, and from wider regulatory and institutional structures. Her research emphasized the importance of managers' readings and experiences of their own and other local retail environments and the need to understand consumer needs and desires (often by conducting research with local consumers), thus abandoning any simple differentiation between producers, consumers, authors and readers of shopping malls.

Ethnographic research has provided different insights than readings of representational retail landscapes into shopping spaces and practices. These have emphasized the sophistication, autonomy and skills of consumers in interpreting consumer landscapes (Jackson and Holbrook, 1995; see Box 3.6). One study traced processes of shopping into neighbourhoods and communities around the London centres of Brent Cross and Wood Green (Miller et al., 1998), exploring how shopping space was understood and interpreted in relation to other shopping spaces. Their study found that shopping practices and pleasures were related not so much to lifestyle choice, or the symbolic construction of commodities, but to the imaginings and fears of places and groups, with commodities a means of negotiating social relations connected to broader structures of family, race and gender.

BOX 3.6 RECONCILING TEXTUAL AND ETHNOGRAPHIC APPROACHES

Textual (semiotic) readings of consumption spaces (and in particular the mall) have come under some criticism (Glennie and Thrift, 1996; Gregson, 1995). This criticism centres on the implicit masculinity of textual readings, the limited ascription of agency to consumers, the (un)representativeness of academics' readings, a bias towards spectacular spaces (often mega-malls), and a lack of context for the construction of social inequalities and

> the material realities of consumption (Gregson, 1995). Such criticisms have largely been advocated by social and cultural geographers drawing on ethnographic traditions emphasizing consumers' experiences and how consumption is enacted, performed, constrained and limited through social practices in particular places.
>
> In defending this method Goss (1999b) argues his readings of the landscape of the mall are not intended to be generalized to groups of consumers; rather they are part of understanding how such built environments work and how 'we' might work against them (Goss, 1993: 18). He advocates that his task as a researcher is not to reveal false consciousness, but to seek to explore the 'redemptive significance' of these spaces, to recognize the latent truth content, and the 'pleasure taken in the play of reality and fantasy' (1999b: 49). Goss believes ethnographic researchers also make readings of texts through interviews, focus groups and observations which represent a privileged reading of the real (in establishing boundaries between reality and fantasy). Goss proposes that critics' concerns about materialism and the surrendering of consumers' wills to the seductions of consciousness industry in these spaces indicate they are also caught up in modern narratives and cultural desires which separate mind from body, culture from nature, present and past and authentic from inauthentic. Most provocatively he argues ethnographic approaches risk banality by reproducing the obvious findings that consumers make their own meanings without engaging in positive and negative critiques of the politics of meaning (1999b: 48).
>
> One means of understanding the debate is to see ethnographic and textual research as performative (in their enactment they bring different ways of being, doing and seeing into being). The two perspectives position 'the consuming subject' differently. In approaches which read landscapes the consuming subject is often framed in relation to the designers, architects and owners of consumption spaces (whether as a passive recipient of visual images or as a resistant shopper) while ethnographic approaches have tended to focus on 'the consuming subject' as a product of the individual and personal (consumer/user) and the social (in relation to material and identity choices and social–spatial practices). Morris (1988: 206–7) argues there is a danger in separating understandings of consumer spaces into the personal (consumer/user) and the professional (designer/manager/owner). Consequently reconciling differences between textual and ethnographic research on the mall must involve a consideration of how subjects and objects of consumption research are made.

The metaphor of consumption as religion and shopping malls as 'cathedrals of consumption' remains attractive because it does point to a partial set of truths (Fiske, 1989). In addition, both those who ascribe to reading landscape as text and those following more ethnographic traditions recognize that architects' and designers' intentions in creating particular meanings in the built environment may be interpreted differently, resisted or even subverted by the people who use these spaces (Urry, 1995; Shields, 1989). For example, those who do not purchase commodities but use the mall as a place to 'hang out', socialize and malinger challenge discourses of consumerism and their linkage to particular spaces (Hopkins, 1991).

Malls are sites of social centrality where people can interact (to varying extents), where new variants of self can be tried out, encountered by others and performed (Glennie and Thrift, 1996). Being in these privately owned 'public' shopping spaces may also be a reflection of other relationships, networks and non-commercial activities. Abaza (2001), for example, notes how shopping malls in South East Asia and Egypt fulfil different functions. A lack of public gardens in Egypt mean malls are spaces for youth to socialize, while in Malaysia malls become a place for the flâneuring of the middle classes, something increasingly difficult in the open air because of traffic congestion. In Box 3.7 Erkip examines how shopping malls have formed an integral part of many Turkish citizens' identity in the 'modern' world.

BOX 3.7 SOCIALITIES AND SUBJECTIVITIES: SHOPPING MALLS IN TURKEY

Spaces of consumption are sites of sociality and subjectivity. Shopping malls are part of the rapid transformation of urban life in Turkey since the 1980s (Erkip, 2003). Feyzan Erkip's (2003) exploration of Bilkent shopping centre in suburban Ankara notes how the created environment of the shopping centre has played a timely role in Turkish urban citizens' search for identity choices. The search for new areas of investment for capitalists, the increasing appearance of women in public space and consumer demand for distinctive and 'foreign' products have facilitated the growth of malls within Turkey. The malling of Turkey has however been uneven, reflecting the social and spatial segregation of the population and the articulation of local–global relations. Malls are more prolific in Istanbul, for example, which has a longer history of global economic and cultural integration than Ankara. In aiming to attract visitors from all sectors of society, Bilkent centre as a new form of public space provides a more heterogeneous and democratic consumption site than previous locally oriented shopping spaces. Surveys of users and focus groups have demonstrated the shopping mall fulfils an important social and leisure function, with many visiting the centre with family and friends and browsing. While the heterogeneous nature of shopping in the mall appears to blur social segregation, research with shoppers reveals attitudes towards others as positively or negatively different remain, with the potential for new arenas of negotiation and conflict, and the emergence of new forms of exclusion (especially for the urban poor). However, Erkip believes the mall also satisifies 'the requirements for a new modernity expressed by all segments of the Turkish society', reflecting new subject positions of urban citizens who use the 'modern and clean environment' of the mall as an alternative to the city centre (2003: 1088–9).

> Erkip's work highlights the ways in which shopping centres are both reflective of and constitutive of new 'modern' subjectivities and socialities. Malls as meaningful shopping spaces must be understood in relation to specific time and space contexts and how these are in turn articulated in relation to other places. Transformation in consumption has occurred as the demands of citizens in urban areas to re-establish modernity have coincided with the efforts of Turkish cities to be more integrated in the global culture/economy (2003: 1090).
>
> The material and symbolic meanings of space and commodities, and the framing of social attitudes, desire, inclusion and exclusivity exist in complex relation between people and places.

How public space is regulated (be this formally or through forms of surveillance) may also limit forms and practices of resistance in place, as non-conforming patrons of these often privately owned 'public' spaces may be removed. Ironically Fiske (1989) suggests that it is through the act of purchase that individuals can exert some means of control over things. Rejection of commodities may be empowering, though in Fiske's opinion 'shopping can never be a radical, subversive act; it can never change the system of a capitalist-consumerist economy' (1989: 27). The popularity of shopping malls and centres can be attributed to their success in managing diversity, reducing the risks of social difference and promoting the virtues of 'familiarity' (Jackson, 1999).

Research by retail geographers (see, for example, Wrigley and Lowe, 1996; 2002) has also extended research on the mall to a consideration of how other shopping spaces are organized to facilitate consumption, and how the construction of these consumer spaces is embedded in political, economic and social relations of production. The construction of contemporary shopping spaces as disorienting landscapes which weaken local distinctiveness as part of homogenizing processes of globalization, for example, has been challenged by work on micro geographies of retailing (Bridge and Dowling, 2001; Crewe and Lowe, 1995) (see also Box 3.6).

Exploring 'alternative' sites of consumption

A difficulty with a schema that posits arcade, department store, mall in an evolutionary trajectory is that each form can tend to be seen as surpassing the last. Arcades, department stores and malls remain features of many contemporary urban spaces and cannot be regarded as historical equivalents in terms of cultural/political/economic significance or consumer patronage (see Figure 3.2).

Attention has also turned to less middle class and more mundane spaces of consumption such as discount warehouses, markets and car boot sales (see Box 3.8). This work challenges the construction of the mall as a space of representation and of lifestyle shopping (see Box 3.6). Geographers concerned with these sites

FIGURE 3.2 Changing significance and use over time is demonstrated by the Queen Victoria Building in Sydney. It was built in 1898 to replace the original city markets, later becoming a concert hall, the city library, and in the 1930s municipal offices. In 1984 it was refurbished and has returned to a retailing function, with more than 200 shops inside

of consumption have also sought to highlight the ways in which consumption is a social, material and bodily experience – recognizing the theatricality, performance, unpredictability, skill, thrift, pleasure and desire of consumption practices and experiences (Gregson and Crewe, 1997a; Crewe and Gregson, 1998). Shopping in these 'alternative' spaces might be risky (see Figure 3.3) (for example, having to negotiate previous bodily presences in clothes, or the prospect of commodities being unusable) but is not necessarily anxiety or identity ridden (Gregson and Crewe, 1997a).

BOX 3.8 ALTERNATIVE SPACES OF CONSUMPTION

Geographers Nicky Gregson and Louise Crewe have argued consumption studies should not neglect material culture or structural social inequalities (for example, differences of gender, class, race, disability and sexuality). Crewe and Gregson (also with Kate Brooks) embarked on a series of research projects focusing on the space of car boot sales and charity shops (see also Chapter 6). This research on second-cycle spaces of consumption involved both participant observation and interviews. Like

> ethnographic work on shopping malls, it reinforced the necessity of understanding consumers' spaces with reference to those who occupy them. Crewe and Gregson (1998) highlighted the transgressive nature of 'alternative'[5] spaces (for example, where people come to play rather than spend at car boot sales), the multiplicity of roles of shoppers (as buyers, voyeurs and entrepreneurs) and how meanings of shopping in space extend across class boundaries (Gregson, 1994; Gregson and Crewe, 1997b). As with Bowler's (1995) research on shopping centre managers, research on car boot sales, charity shops, second-hand children's clothing stalls and catalogue shopping (Clarke, 1998; 2000) has demonstrated the indistinct nature of divisions between consumers and producers of consumption and the interconnections between different types of shopping experience and practice.

FIGURE 3.3 Shopping in alternative consumption spaces can be a risky business, as the name of this shop attests!

In recent years a distinction has been made between retail geographies and geographies of shopping. The latter are concerned with explicating how shopping is talked about, understood and practised by consumers (for example, in terms of necessity or choice) rather than with retail spaces *per se* (Gregson et al., 2002a). This work is beginning to explore how practices of consumption are linked to spatialities. Thus second-hand charity shop purchasing or high street shopping, for example, have been shown to be constitutive of other practices, imaginings and discourses which resonate across places and scales, such as moving to and from cycles of first-hand shopping (see, for example, Miller et al., 1998; Crewe and Gregson, 1998), with space 'both limiting and enabling possibilities for particular subjects' (Gregson et al., 2002a: 615).

Though geographic literature on consumption spaces tends to have been dominated by shopping or retail geographies, other sites and forms of consumption have been examined. The next section considers the home as a site of consumption, a site in which the workings of power are manifest within, across and beyond its boundaries.

Home: Scaling Public/Private Spheres

'The place where we live is one of the key locales which shapes our sense of place and enables us to develop our sense of who we are' (Perkins and Thorns, 1999: 124).[6] Homes are sites in which meanings are constructed; they may be sites of pleasure, fear, relaxation and work, security and insecurity for different inhabitants (even altering at various times in different spaces of the home). Concepts of home are influenced both by the social-political-economic contexts in which houses are created and by the representational landscapes and discourses of home embodied in legislation, television, media, architects' plans, art, literature, advertising, real estate sales and so on.

Home has been and continues to be an important site of scale construction operating to influence nations, regions and communities (Marston, 2000). Towards the end of the nineteenth century the dominant hegemony which had located women in the private sphere and men in the public sphere in many Western countries, began to be eroded. For nineteenth century middle class women, structures of social reproduction and consumption began to transform the home into a form of public space. The creation of women's movements, the application of Taylorist principles of scientific management to domestic work and technological advancements all contributed to home-making as a public function, with women as productive state citizens and with activities within the home shaping democratic nation and community.

Homes are also constituted through the enactment of power via household members' roles and relationships. The complex ideological and material construction of home involves different understandings and norms about appropriate roles and behaviours, characterized by aged, gendered, sexed, classed and racialized roles and obligations (Bowlby et al., 1997). Homes provide a setting for the contexts people create with commodities (Sack, 1992) and the material cultures associated with them which are both an appropriation and a representation of the world (Miller, 2001a). Purchasing and using commodities, furnishings, décor, leisure and housework equipment can be powerful household practices, complicit in the identity construction and politics of households (Bowlby et al., 1997). Discourses of heterosexuality and the home, for example, both pose opportunities for and put constraints on the construction of lesbian identities through material consumption and display (Johnston and Valentine, 1995).

Despite the blurring of the division between public and private spheres in contemporary Western societies and challenges to the cult of domesticity, women still tend to have the main responsibility for domestic work within the home (L'Orange Fürst, 1997). 'Home-making', particularly in the form of furnishing and decorating, is one of the key elements available to Western woman for putting her 'stamp' on her surroundings (Goodall, 1991: 275). Suburbanization and processes of gentrification in many Western countries have been accompanied by DIY ('do-it-yourself' home based renovation) (Redfern, 1997). While this is the premise of both men and women, for much of the twentieth century DIY has been associated with men's home based work (Clarke, 2001), thus superficially establishing a difference between 'home-making' (a feminine function often assumed to comprise consumption and an ethic of care) and 'making home' (a masculine, productive and economically value producing activity).

Box 3.9 examines the concept of domestic technologies and tools (as non-domestic technologies) in constructing masculine and feminine subjectivities, practices and places. Comparisons between the US and NZ experiences illustrate how subjectivities are constituted in similar ways but differ according to social-spatial context (in NZ's case masculinity is articulated more strongly through discourses of rurality and frontier culture).

BOX 3.9 MAKING HOME AND HOME-MAKING: TOOLS OF GENDER?

Gelber (2000) argues 'Mr Fixit' put in his first formal appearance around the turn of the twentieth century in the United States. Previously, middle class Victorian households had hired professionals to do home repairs and improvements, with men finding spaces other than the home (such as fraternities, clubs) in which their masculinity might be reaffirmed. In common with many nations in the 'West', in the USA suburbanization, growing home ownership and greater availability were critical components in the shift from 'restrained and distant father to engaged suburban Dad' (2000: 71). Over the twentieth century the increasingly sexually integrated workplace and the growth of white collar employment also produced sites of more ambiguous masculinity. One way heterosexual masculinity could be reasserted was through 'direct male control of the physical environment and through the use of heavy tools in a way that evoked pre-industrial manual competence' (2000: 71). DIY projects involving household construction, repair and maintenance were free 'from any hint of gender-role comprise' and by 1950 the ability to engage in home DIY was virtually a requirement of US masculinity (2000: 71).

FIGURE 3.4   Domestic masculinities in New Zealand still appear to be associated with the myth of the 'good Kiwi bloke' as DIY expert, often having his own space in which to be productive

The emergence of DIY in New Zealand/Aotearoa homes over the nineteenth and twentieth centuries is also suggestive of a type of masculinity, but one which is more clearly framed as rural and productive (Berg, 1994). Urban spaces in New Zealand are seen as feminine consumptive spaces and as spaces of soft men.[7] In New Zealand hegemonic Pakeha masculinity emerged out of nineteenth century colonial life and the predominantly male communities of rural/frontier regions. It was in these frontier spaces that resourcefulness, practicality, versatility, physical skill, independence and ingenuity became traits of what came to be known as the 'good Kiwi bloke' – masculine and heterosexual – who in the 1920s became associated with the good 'family man' (Phillips, 1996). The stereotypical family man of the 1950s to 1970s was proficient at fixing things and in gardening. As in America, the home (as opposed to the pub, the bush and the rugby field) provided a problematic space in which to find reassurance of one's masculinity:

> But what if he spent the weekend at home, then what? Was not the home the woman's world? The man's response was to cordon off from the domestic environment certain exclusive male territories. Fences of sexual segregation were erected at home. The man would not cook unless it was over a campfire; he would not clean unless it was the car; he was prepared to garden so long as it was always vegetables and not the herbaceous border; he was prepared to mend things so long as it was a washer and not socks; and he was ready to cut wood. The jobs that were acceptable were those that generally involved heavy physical work or mechanical skills – outside tasks which allowed him to relive the fantasy of the pioneering life. (1996: 243)
>
> Thus weekends in New Zealand 'were often punctuated with the sounds of the home-handyman's tools, the growl of the motor-mower and chainsaw and the whine of the power drill, as New Zealand men demonstrated their ingenuity and commitment to a culture of "do-it-yourself", based around fixing and producing rather than artistic creativity' (Perkins and Thorns, 2001: 44). In contemporary New Zealand, DIY renovation continues to be an important home-based pursuit (see Figure 3.4) for both men and women (Perkins and Thorns, 1999). While the resourceful hyper-masculine Kiwi bloke remains an iconic construct (particularly in advertising), social and spatial change both within and outside the home has seen a much broader range of masculine (and feminine) subjectivities emerge in recent years.

Thus consumption of machines, tools and technologies can have an important role in the gendering of the space in and around the home and in the constitution of feminine and masculine subjectivities. While domestic technologies (microwaves, vacuum cleaners etc.) are used by men, domestic technologies are not generally considered 'men's machines'. The domestic sphere is generally regarded as non-technological, undermining the home as a place of significant activity and influence (Cockburn, 1997). Scanlon (2000) argues males' use of the electric carving knife and barbeque technologies allows them to participate in household culture without compromising their identity. The discussion of masculinity in the US and New Zealand has been linked with objects labelled 'tools' rather than 'appliances', the latter being more clearly linked to the domestic sphere (traditionally the woman's world). However, gender and technology are not fixed entities but are part of wider networks and systems of production and consumption, cross-cut by structures, politics and practices of class, race and sexuality.

Conceptions of home as private feminine space and of outside home as public masculine space have also been challenged in academic discourse (Stevenson et al., 2000). In addition a danger of studies which assume domestic work is oppressive is that they may portray women and men as belonging to fixed and unchanging identity groups (Cameron, 1998). Thus a potential difficulty of the conception of 'men as builder and women as decorator of public space' outlined in

Box 3.9 is that it tends to position men and women as universal occupiers of space without suggesting the possibility of or acknowledging the existence of alternative forms of masculinity, femininity and their performance and negotiation in the home space. Cameron (1998) argues for a more fluid understanding of politics and gender, demonstrating through interviews with women how identities are transformed as they move in and out of subject positions of worker, wife and mother.

The case study of DIY also neglects another aspect of consumption that geographical studies have attempted to address, which is the specific meanings, practices and discourses associated with consuming tools and domestic technologies in the home. Ethnographic work at a number of sites (Miller and Slater, 2000; Clarke, 2001) has begun to unpack how objects become incorporated into individuals' lives in particular settings, and how the consumption of objects is implicated in the production of scale and cultural politics. Research with Internet users, for example, is not only changing the way in which consumption sites and spatialities of consumption might be perceived, but also providing exciting insights into how place is implicated in the production of subjectivities.

Geographies of Cyberspace

Information and communication technologies increase the permeability of home boundaries with regard to the relative ease with which information, behaviour and presence flow – bringing the world outside in, and taking the home 'out' (Shapiro, 1998). Trends towards home based consumption via the telephone, the Internet, cable networks, credit cards, electronic points of sale, loyalty 'smart' cards, 'cybercash' and home based shopping and banking facilities are in Graham's opinion creating systems of surveillance which 'precisely monitor, in real-time, the consumption patterns of households' (1999: 137). Such technologies have grown rapidly over the last few decades and have been complicit in the creation of new spaces, ways of consuming and means of communicating with others.

The Internet and cyberspace: inclusions and exclusions

The Internet, in linking people through new spaces, is 'changing the way we think, the nature of our being in the world and our identity' (Turkle, 2002: 456). The boundaries between public and private may be altered by the availability of personal material in the public domain, the possibilities of electronic surveillance and the opening up of one's world to collective consumption in the form of web pages. Online and virtual gaming, trading pictures, tourism and place marketing, chatrooms, and e-shopping for commodities (see Box 1.1) and services all provide media for altering existing and constructing new spatialities, socialities and subjectivities. While some of the sites of the Internet are places of work 'they are also spaces of consumption (the space itself is consumed) and many are spaces of pure consumption; they exist only to be consumed' (Dodge and Kitchin, 2000: 30).

The Internet is a global network of computers that are linked together by telecommunications technologies with 'each linked computer nesting in a hierarchy of networks from its local area to its service provider to regional, national and international telecommunications networks' (2000: 2). The most common form of Internet activity is sending e-mail, but surfing 'the web' is increasingly popular. The World Wide Web, which consists of multimedia data stored as hypermedia documents linked to other pages of information, has become the fastest growing medium in history.

Cyberspace is 'navigable space', the digital space of networked computers accessible from computer consoles which exists within the infrastructure of cyberspace, the hardware of the digital world (2000: 1). Cyberspace can be viewed as a transformative agent, playing a role in the production of space and scale, and altering relationships between people, places and objects. As mentioned in Chapter 1, cyberspace may have induced space–time compression, but the annihilation of space or place has not resulted (Walmsley, 2000). The uneven diffusion of and access to the Internet for consumers – a result of variance in the availability of bandwidths, telecommunications technology and hardware, skills and training – have meant new forms of unevenly developed social relations and possibilities of new forms of social exclusion (Dodge and Kitchin, 2000). The Internet is dominated by Western or developed countries, particularly the United States of America with 65 per cent of all website hosts in January 2000, followed by Canada (9.5 per cent), Japan (3.6 per cent) and the UK (3.3 per cent) (Jordan, 2001: 3). In addition developed countries account for the greater percentage of traffic and hosts in other regions of the world, with 80 per cent of website hosts using English as a first language (2001: 4).

Cyberspace is not simply a space of infrastructure, but is a space of social relations which can play an important role in constituting consumption and identity politics and practices, and in the associated formation of social networks and power geometries. While the net may be associated with unfettered information and freedom, access to this is not unmediated (Rule, 1999). It has been increasingly privatized and dominated by online service providers selling value-added services (Kitchin, 1998). Search engines provide particular information possibilities (which may be influenced by commercial objectives or advertising revenues) and many websites deposit cookies on personal computers which provide identifying data on consumers and preferences for return visits (Rule, 1999). It is also a site of risk, with the possibility of surveillance and personal intrusion by governments, institutions and individuals through monitoring of e-mail conversation, the increased collection and storage of personal data both online and on digital databases, the accessing of individual terminals through publicly broadcast IP (Internet Protocol) numbers, by hacking, and also the risk of virus infection (Graham, 1998a; 1998b).

Cyberspace is often but wrongly presented as an impoverished simulation (copy) of real-world contexts or as a hyper-realization of the real (Doel and Clarke, 1999).[8] Real and virtual worlds are actually interconnected and inseparable, with cyberspace being a (personal) part of (rather than apart from) many individuals' lives (Miller and Slater, 2000). E-mails, for example, can strengthen rather than diminish social relationships in 'real' space. The production and consumption of cyberspace occurs in multiple contexts outside the net. Many 'virtual spaces' are formed through their associations with real-world contexts such as tourist and place promotion websites, virtual retail stores, and government, NGO, community action or company websites (Graham, 1998a), and real-world practices and terms (such as 'site', 'travel', 'surf' and so on) are used to describe the Internet (Dodge and Kitchin, 2000). Consumers of cyberspace are simultaneously producers of socialities and particular identities (as avatars in simulated worlds, game players, participants in chatrooms, discussion groups etc.), with these productions both located in and articulating relations of physical space. Such productions are also organized and structured and related to 'real-world' social and spatial relationships: computer users engaging in 'free' exchange of sex-pics, for example, produce structures which valorize and regulate rules of exchange and establish particular moralities (Slater, 2000).

The Internet and other technologies of cyberspace (including Internet capable mobile phones) also potentially increase the encounter between home, work, leisure and other spaces and the wider world. Home pages, for example, open one's world up for collective consumption by unknown others.[9] Like 'picturing' as a practice in the consumption of tourism (Crang, 1997), the human interactions in and construction of cyberspace may provide a means of examining the ways 'images are embedded in time and space' and 'how they place with space and time' (1997: 371) . However, we cannot expect that the Internet will impact on, be consumed in or be incorporated into individuals' lives in definable ways (Graham, 1998b). Kitchin (1998) suggests that it is only recently that geographers have begun to address the political, social and cultural implications of cyberspace. While cyberspace has facilitated the development of new spatialities and socialities in online environments, it has also been complicit in the emergence of off-line social and spatial relationships (for example, in the organization of raves, meetings, protests, local exchange trading schemes). New spaces have emerged, with Internet cafés for example, creating fascinating insights into how virtual and real worlds intersect (Miller and Slater, 2000; Wakeford, 1999). The inseparability of real and virtual spaces is explored Box 3.10.

BOX 3.10 CONSUMING CYBERSPACE: LINKING REAL AND VIRTUAL WORLDS

Though the physicality of cyberspace cannot be humanly habited in the same way as 'real' space, cyberspace is also a space in which the symbolic and the material are mediated (Turkle, 2002). Research into how people have used the Internet has shown that cyberspace is not a disembodied space but a product of 'social, institutional, political and economic processes that shape spatial arrangements and interactions both on- and off-line' (Dodge and Kitchin, 2000: 28). Miller and Slater's (2000) ethnographic study examined Trinidadians' affinity with the Internet. They demonstrated the Internet was complicit in the construction of what it means to be a Trinidadian and that, in turn, being Trinidadian meant certain affinities were associated with the Internet as an object of material culture. They concluded that cyberspace does not exist in a placeless vacuum but is personalized and made meaningful in place.

The Internet is also shaped through place rooted cultures (Holloway and Valentine, 2001a). The Internet was initially a product of the Cold War, emerging as a solution to the potential disruption of US military and other communications in the event of a nuclear war (Hurwitz, 1999). Given the continued dominance of American hosts on the Internet, and the disproportionate embedding of Internet technology in the USA, Holloway and Valentine argue the place rooted cultures which dominate online space are more likely to be American. In a study which explored British children's (age 11–16) use of the Internet at school and at home, Holloway and Valentine (2001a) clearly demonstrated the mutually constitutive nature of real and virtual space. Most of the children used the Internet to further their general interests. Many of these Internet encounters were Americanized, but not simply because of the hosting of sites in cyberspace; it was also because their offline interests were already formed around American film, TV and consumer cultures. Some children also deliberately sought to consume non-American sites on the net, again as a response to their place based positionality and subjectivity. Holloway and Valentine's research, and that of Miller and Slater (2000) on Trinidad, explores the relationality of real and virtual spaces and touches upon how cyberspace is constituted and experienced in specific geographical contexts. There remains much to be understood about how such technologies and spaces are interpreted, altered and incorporated into the practices and discourses of everyday life and in turn about how everyday spatialities, subjectivities and socialities are manifest in cyberspace.

Geographic studies of cyberspace have emphasized the sphere of production rather than consumption *per se*, centring upon the information economy and its effect on employment patterns, economic performance and urban–regional development (Kitchin, 1998: 388). However, Wrigley et al.'s (2002) discussion of e-commerce has identified the significance of cyberspace as a site of consumption. Wrigley et al. outline the emergence of spatialities and socialities linking real and virtual retail space. These include disintermediation (the direct linking of consumers with producers), the creation of novel ways in which potential consumers might engage with commodities (such as videos, concerts, sound bytes), and the emergence of online informediaries who operate between the customer and the retailer as providers of information to enable online consumers to make informed purchasing decisions. The form of the e-retailer (a pure e-tailer, or a store based retailer with an e-commerce division) and the commodities can also facilitate or constrain the possibilities for consumption online. Virtual or electronic products such as MP3 music files, or e-cards which can be selected, experienced, transported and consumed online, present vastly different possibilities for consumption than those commodities where consumers' experiential possibilities are limited.

The idea that 'computers don't just do things for us, they do things to us' (Turkle, 2002: 46) is a provocative one. Much has been written about how the Internet allows people to adopt different roles and personas online (Robson, 1998), but Turkle's quote also alludes to the ways in which technologies and their consumption might operate in social-spatial contexts. In Box 3.9 on making home, 'tools' were discussed as objects which may be understood, used and objectified differently, but the tools themselves were seen as stable entities, unchanged through the consuming relationship in which they were used. Research on cyberspace has demonstrated that technologies can have a role in influencing the associations between human and non-human actants (see Chapter 5 for a discussion of actor network approaches). Properties and meanings of cyberspace also emerge in offline communities of practice (such as through home and school) and through the linkages between the two which produce spatialized discourses of childhood, danger, risk, technical competence and gender (Holloway and Valentine, 2001b).

Cyberspace presents possibilities for the creation of new spaces of disintermediation where consumers and producers are linked and commodities defetishized (as in buying ethically produced food: see Box 1.6), creating a new progressive politics (and geometry) of connection (Miller, 2003). Thus research on the Internet and cyberspace has the potential to challenge existing notions of 'the consumer', what consumption is, and the places and spaces through which it is manifest.

Consuming Spaces

This chapter has examined the significance of consumption in making place and scale. Mort (1998) highlights the danger of positioning consumers in relation to single or isolated series of consumption goods, suggesting this makes it difficult to follow the more dispersed and extended networks through which commodities are circulated and the cultural rituals through which goods are inserted into the fabric of everyday life. Similarly, positioning consumers in places that are framed as inclusive spaces (be it mall, market or theme park) may ignore the ways in which consuming experiences, practices and representations are embedded in the production of multiple spaces and scales. Within consumption studies, geographers have focused on bounded spaces such as the department store, the mall on shopping centre, the car boot sale, charity shops and the home.[10] However, much of their work (particularly that of the ethnographic tradition) promotes the sort of understanding Mort advocates, demonstrating the active creation, fluidity and relationality of sites of consumption. Geographers have explored how change and transformation of places at one site (be it the home, the body or the market) are intimately connected to other places and scales, and to particular constellations of social relations manifest in symbolic and material forms. The literature outlined here has also implicitly touched upon the workings of power operating through disciplining practices and discourses of surveillance, control, inclusion and exclusion (whether expressed as the surveillance of mall spaces or parental control over access to the net). Crang (2002) calls for more studies of the interstitial spaces of everyday life, that is, spaces of circulation, mobility and ephemerality (such as airports, motorways, cable networks). Extending the work already done on more bounded spaces in consumption, geographers are consequently beginning to provide insights into the construction and socialities of such important but oft neglected spaces and places. The discussion of domestic technologies and the Internet has focused on how people take and make meaning through consumption in place and how this relates to identity formation. In the next chapter, these ideas are examined more explicitly, emphasizing connections between place, people, commodities and the consuming practices they engage in.

FURTHER READING

Crewe, L. and Gregson, N. (1998) 'Tales of the unexpected: exploring car boot sales as marginal spaces of contemporary consumption', *Transactions of the Institute of British Geographers,* NS 23 (1): 39–53.

Dodge, M. and Kitchin, R. (2000) *Mapping Cyberspace.* Routledge: New York.

Goss, J. (1999b) 'Once-upon-a-time in the commodity world: An unofficial guide to the mall of America', *Annals of the Association of American Geographers,* 89 (1): 45–75.

Holloway, S.L. and Valentine, G. (2001) 'Placing cyberspace: processes of Americanization in British children's use of the Internet', *Area,* 33 (2): 153–60.

Marston, S.A. (2000) 'The social construction of scale', *Progress in Human Geography,* 24 (2): 219–42.

Miller, D. and Slater, D. (2000) *The Internet: an Ethnographic Approach.* Oxford: Berg.

Miller, D., Jackson, P., Thrift, N., Holbrook, B. and Rowlands, M. (1998) *Shopping, Place and Identity.* London: Routledge.

Mort, F. (1998) 'Cityscapes: consumption, masculinities and the mapping of London since 1950', *Urban Studies,* 35 (5/6): 889–907.

NOTES

1 See, for example, the forum on 'Place Matters' in the *Annals of the Association of American Geographers* (91(4), 2001), and the debate over the social construction of scale in *Progress in Human Geography* (Marston, 2000; Marston and Smith, 2001).

2   Ritzer (1993) states rationalization of spaces has occurred as places have become more McDonaldized. This involves systems of production, distribution and consumption which are characterized by efficiency, predictability, calculability and the replacement of human with non-human technology.
3   Rappaport (2000: 39) suggests the concept of the flâneur makes women the object of the masculine gaze. Flâneurs could both window shop and women shop, consuming with their eyes and feet. Others caution about the use of this term in contemporary consumption spaces, suggesting the 'mall-walking dope is more akin to a baudaud (a passive intoxicated observer) than a flâneur (an active joyful watcher)' (Woodward et al., 2000: 352).
4   The terms 'mall' and 'centre' appear to be used interchangeably in the literature. However, despite the rise of planned big box, power or warehousing shopping centres and retail parks where people walk or drive between shopping units, and the existence of centrally planned and managed speciality outdoor shopping centres, the term 'shopping centre' and 'shopping mall' most usually applies to enclosed shopping spaces with controlled physical environments, normally owned and/or managed by a single operator.
5   Defining such spaces as alternative can actually reprioritize shopping malls and spectacular consumption spaces, positioning sites of first-cycle consumption as 'more significant'. It also tends to suggest the two operate independently, whereas Gregson's own research has clearly demonstrated their inseparability and relationality.
6   The perspective of homes here is largely drawn on the 'white, Western ideology of the home', one which prioritizes a physical entity as the site of dwelling and a set of social, economic and sexual relations (Bowlby et al., 1997: 344).
7   The notion of urban space as feminine also exists in contrast to much writing which positions public space as masculine (Berg and Kearns, 1996: 104).
8   Castells (1996) suggests that the virtual can have a reality of its own through the concept of 'real virtualities', in which the virtual becomes an alternative space rather than simply a form of representation (see Crang et al., 1999: 6–8, for a fuller discussion). Miller and Slater (2000: 6) suggest the concept is misplaced in that it re-cites a division between the real and the virtual. They suggest that virtuality as 'the capacity of communicative technologies to constitute rather than mediate realities and to constitute relatively bounded spheres of interaction' is neither new nor specific to the Internet (for example, newspapers, government policy documents, television etc. also reflect particular social-spatial imaginaries).
9   Technologies of cyberspace may be productive of new power relations, evoking possibilities for challenging, reappropriating and creating social-spatial relations through what Haraway (1991) terms 'cyborg politics' (formed out of the hybridity of technological and human embodiment, the blurring and re-creation of the people–machine interface).
10 Though not discussed in this chapter, a significant amount of literature has also emerged on cities as consumption spaces. Much of this focuses on discussing cities as a manifestation of consumption in contemporary society and on services as cultural economy (Clarke and Bradford, 1998; Paolucci, 2001; Scott, 2000; Zukin, 1998; and see also the range of articles in *Urban Studies,* 35 (5/6), 1998).

4

Identities

Consumption is a medium through which people can create and signify their identities. Theorizations of postmodern society suggest consumers are enmeshed in a world of commodities in which decisions about who one is and how one should be (in what spaces) are becoming increasingly complex. While consumption can play a significant role in projects of self-fashioning, geographers have shown discourses and practices of consumption are not solely about identity formation. Consuming involves practical and bodily experiences which may be about mundane acts and provisioning, security and sociality as much as individualistic lifestyle choices. Discourses and practices of consumption also locate bodies in particular spaces, emplacing identities. Places in turn influence processes of embodiment, influencing the rituals, practices and consumption through which they are made meaningful.

The chapter begins by exploring processes of identity formation, and then moves on to examine connections between consumption, embodiment and emplacement by exploring the concept of performativity. In the last section of the chapter, geographies of food are shown to provide critical insights into how relationships between subjectivity and space are manifest.

Consuming Identities and the Postmodern Condition

Identity formation is concerned with one's subjective sense(s) of self and is about one's sense of being and processes of 'becoming' and belonging in place.[1] People have multiple and fluid identities which are formed not only as a reflexive positioning of self, but as a process which occurs in relation to others who are distant from the self (Rodaway, 1995). Processes of identity formation involve creating meaning in the space of one's physical body, which also involves a consideration of how our bodies are interpreted and located in wider discursive and material contexts.

The postmodern condition is viewed as facilitating a crisis of identity where there is a disjuncture between one's experience in social contexts and one's meanings of self and others (Pred, 1996). Identities in a postmodern epoch are viewed as transitory, less attached to enduring social structures and relations (such as in relation to paid employment, or to a stage in one's life course). An aestheticization or stylization of life ensues, with consumption playing a more significant role in the regulation and construction of identities. The multitude of fragmented identities presented and represented to consumers through commercial culture is believed to cause insecurity. Consumption becomes driven by fantasies which fuel desires which cannot be satiated, resulting in a constant search for commodities which enable the self-fashioning of a lifestyle and an identity space (Friedman, 1994).

The postmodern schema advocates a necessary connection between identity formation and purchase of commodities, yet the relationship between identity formation and consumption is complex, multifaceted, shifting and even contradictory.[2] The material aspects of commodities and their use values tend to be obscured behind narratives which claim identity and symbolic value are of primary importance. The postmodern narrative also disregards the historical importance of consumption in the construction of identity (see Chapter 2). Mackay (1997) indicates there are two ways of conceptualizing relationships between consumption and identity: consuming to become, and consuming according to who we are.

We consume to become who we are?

In this schema individuals create, affirm and contest social identities through consumption practices. Taken to the extreme, this perspective is usually associated with the purchase of commodities and is enshrined in notions of 'lifestyle shopping' (Shields, 1992a) where individual consumption is predicated on acquiring the qualities and identities associated with commodities and particular consuming practices. Such a view draws on Baudrillard's (1981; 1988) writings, in which consumers disappear in an overwhelming commodity world of signs and simulations, and consumption becomes an end product of consumer manipulation, need and utility. As Goss explains, 'It seems we are all consumers now. But perhaps we are not even really that, and have ourselves become objects of consumption' (1999a: 114–15). Consumer identities and practices are thus sold as products and objects which may possess subjects through their characteristics. Rowlands argues that this has occurred in Cameroon, for example, where Western commodities have 'become the touchstone for the production of selfhood' (1994: 150).

A view of individuals as consumers who are engaged in the construction of self by purchasing commodities which will provide distinction results in a somewhat superficial and one-sided view of consuming subjects (Falk and Campbell, 1997). People are reduced to shoppers, consumers and performers of 'social identities', neglecting the spatiality and relatedness of experiential bodily practices and processes of cognitive self-reflexivity. Yet this 'relatedness' may be critical in choices not to buy or to consume, for example, decisions about whether a commodity is for me, imagining whether one is or could be 'like that?' (Falk and Campbell, 1997).

The tendency to reduce individuals to lifestyle shoppers, actively choosing commodities and identities, is challenged by geographical research with shoppers (Jackson, 1999). Shopping is often routine and mundane work and identity is frequently articulated in relational rather than positional terms (such as through notions of the family, or racialized understanding of place). Practices of consumption may also intersect with and partly acquire their meanings from other spheres of social life, other practices, practical knowledges, popular experiences and lifestyles (Mort, 1988).

We consume according to who we are?

Rather than practices which emerge from what subjects consciously or unconsciously 'desire to become', a second predominant sense in which consumption has been used is that it occurs because of who we are. Here consumption becomes the articulation of one's sense of positionality as 'more and more people define themselves according to their style of living' (Gottdiener, 2000: 3). Thus the consumption of goods constitutes one's expression of taste, income, employment, gender, position and so forth.

Empirical research has indicated that class, gender, cultural, generational and family identities influence the construction of personal identities (Lunt and Livingstone, 1992). Saunders (1989) concluded that social structures influenced consumption, arguing from his study of housing classes and consumption cleavages that consumption had actually replaced production as the major source of class differentiation in contemporary society. Similarly Crompton (1996) argues class continues to shape life chances and attitudes in contemporary society.

Numerous studies have also pointed to the ways in which consumption practices may be constituted in relation to but not necessarily as a consequence of particular discursive social and materially based categories. Pain et al. (2000), for example, have investigated the intersection of class and old age in leisure practices and sites; McRobbie (1993) has examined consuming cultures associated with teenagers; and Binnie (1995) has explored the 'pink pound' and gay consumption. Much of this later work has interpreted 'social groups' in a non-essentialist and non-homogeneous manner, making consumption practice and sites and/or the discourses surrounding them the focus, rather than an essentialist and homogeneous typology of consumers such as women, youth, the old and so on. Bourdieu's (1984) writings on the cultural construction of consumption have also been influential in conceptualizing the relationships between commodities, social structures and identity formation (Box 4.1).

> **BOX 4.1 BOURDIEU: CULTURAL CAPITAL, DISTINCTION AND IDENTITY FORMATION**
>
> Like Veblen (see Chapter 2), Pierre Bourdieu focuses on the symbolic or 'identity value' of commodities to demonstrate how difference is constructed through the consumption of goods. Bourdieu explored how one's locatedness in society (in this case class based French society in the 1960s) could result in particular consumption practices and processes of self-identification. He suggested there was a relationship between group identity/membership and consumption practices, arguing that social differentiation and distinction were based less on wealth itself, and more on the ability of different classes to display wealth and invest cultural value in symbolic goods (Miller, 1987: 148). The dominant fraction of the dominant class in society was generally seen as having social status and prestige as the owner of cultural (or symbolic) capital. Cultural capital is conferred through cultural practices which involve the implementation of taste in or judgement of commodities and discrimination of their 'signs' (such as knowledge of what constitutes a 'good' bottle of wine and the practices through which it might be acquired, used and appreciated). Bourdieu viewed education as a principal means of discriminating between social groupings, something usually acquired via economic capital.
>
> Bourdieu (1984) used the term 'habitus' to encapsulate the socialization of the desires of the class condition, which are also inscribed on the individual's body. Bourdieu viewed habitus as the local configuration of practice, a set of enduring and often unconscious dispositions, feelings and preferences which provide a framework for establishing 'taste' and which guide behaviour. Thus habitus involves the contextualized space of lifestyles (Bourdieu, 1984), the creation of both cultural domains and the relationships with others who share the same prejudices and preferences about the nature of things (Miller, 1987: 153). While Bourdieu saw habitus as structuring the individual's dispositions, reflexivity could enable one to interrupt its deterministic tendencies (Ritzer et al., 2000). Bourdieu's analysis has been criticized for being moralistic, with a working class driven by sensual, physical and immediate desires, and an educated upper class which is able to 'resist' and cultivate a distance from these desires (Miller, 1987: 151); and for being overly deterministic, with the assumption that people's habitus is already established prior to consumption (du Gay et al., 1997). Nevertheless Bourdieu's writings have made an important contribution to consumption research, linking subject positions with material and symbolic meanings of commodities. In highlighting the ways in which the two spheres are linked in social contexts, Bourdieu explored a theme taken up much more explicitly by geographers examining relationships between subjects, commodities, society and space.

In both the 'consume to become' and the 'consuming according to who we' schemata it is easy to posit the subject as an object of consumption, as one who (must) purchase identities to establish a coherent sense of self, or as consumers whose (classed) habitus is the source of all consumption. Both perspectives offer partial understandings of how consuming subjects may be constituted. Identities may be as much about belonging and sociality, practical knowledges and provisioning, as they are about representation, distinction and individuality. Binnie (1995: 187) states, for example, that consumption can be both an assertion of gay economic power and a response to ameliorate the powerlessness felt in gay people's lives. Gillian Swanson's (1995) writing on the relationship of women, sexualities and city spaces illustrates well the inseparability and the immutability of these constructions. She believes consumption and identities are fluid, neither pluralistic nor located within a static social grouping. Swanson argues that the concept of 'affinity' (that is, patterns which emerge from particular relationships of similarity and difference) can be used to enable meaningful and non-essentialist understandings of women's varied encounters with the city.

An emphasis on purchase of commodities for identity purposes (implicit in postmodern narratives) renders invisible the diverse ways in which subjectivities are affirmed and contested through consumption practices, rituals and discourses (Jackson and Thrift, 1995). Identities are also attached to bodies, making a consideration of bodies important to understanding relationships between consumption and processes of subject formation.

Body Matters

In recent years geographers have begun to problematize bodies and issues of corporeality. This work has viewed bodies as sites, as places of location which are both metaphorical and material. Bodies are both a source and a repository of power and are made meaningful through cultural politics, a politics which 'emplaces' them in varying spatial and temporal contexts. Boundaries between subjects and objects, self and others result, the spatiality of which is intimately connected to desire, disgust and fascination at things outside the self (Pile, 1996).

Physical bodies are always social bodies because they are produced through social relations, trained and disciplined (Benson, 1997). Processes of identity formation occur through the body, where identities may be 'enacted, negotiated or subverted through bodily practice' (1997: 159).

Longhurst (1997: 488–9) categorizes three main approaches geographers have used to understanding the body. The first approach places an emphasis on pre-discursive 'lived bodies' through phenomenological approaches employed in the work of humanistic geographers. The second encompasses a psychoanalytic approach, drawing on the work of Freud and Lacan to understand how sexed identities are created, and primarily associated with French feminists Irigaray, Kristeva and Cixous. Longhurst (1997) critiques the third (and overlapping) approach which revolves around social construction for its tendency to reduce the body to a system of signification, but nevertheless suggests it has much to offer geographers because its proponents argue bodies and places cannot be understood independently of each other.

It is this third 'constructionist' approach which appears to have been extensively utilized in the work of geographers interested in issues of identity and consumption (for example, Aitchison, 1999; Binnie, 1995; Bell and Valentine, 1997; Jackson, 1989). This approach treats the body as a cultural construct and a constantly reworked surface of inscription (Grosz, 1994). The ideas of Elizabeth Grosz (1994) and Judith Butler (1990) have been drawn on to understand how bodies are constructed by others and self to produce identities which are constitutive in part through particular organizations of space (Rose, 1995). The complex interaction between the social and the physical body provides the context for processes of identity formation (Box 4.2). Identities are constructed in the ideological realm which positions bodies in particular discursive spaces which are both self-nominated (embodied) and externally imposed (emplaced).

BOX 4.2 GEOGRAPHIES OF THE BODY MATTER: THE WARDROBE MOMENT

The 'wardrobe moment' encapsulates the complex task of dressing and the dilemma facing many women in the West, where a woman stands in front of her wardrobe and wonders what to wear (Banim et al., 2001: 1). Her choice may be informed by consideration of what fits, what's clean, what matches, how she feels, where she is going and what she intends on doing. Such considerations are not simply self-imposed but involve encountering risk and other places, for example the risk that when one's body enters public and/or social space the appropriate choice has been made. It also involves issues of power, confronting the dominant ideologies which are centred upon appropriate clothes and the ways in which women (and men) should appear (2001: 6). However, people may not simply transfer the meanings embodied in clothes to their own bodies, but may use clothes to reappropriate, transform or subvert hegemonic modes of presentation and identification in space. Bodies (clothed and unclothed) thus become 'maps of meaning' – maps of the relations between power and identity, the place where social practices happen and the place where tensions between creativity, self-expression, anxiety and dissatisfaction are articulated in relation to other places and contexts (Jackson, 1989; Rose, 1993).

The inscription of bodies is of relevance to geographers interested in consumption. Bell and Valentine state that 'Discourses in the media, fashion, industry, medicine and consumer culture map our bodily needs, pleasures, possibilities and limitations. These cartographies produce geographically and historically specific "norms", within which we locate, evaluate and understand our bodies'

(1997: 26, citing Gamman and Makinen, 1994). Research on bodies out of place demonstrates how 'bodies matter' and how 'body matter' is important not just to the embodiment and emplacement of identity, but also to the operation of power in relationships between consumption, identity and place.

Bodies out of place

Consuming and processes of identity formation occur within 'moral' discourses which endeavour to define appropriate consuming practices, relationships, objects of consumption and even the places in which these should occur. Places are critical to how bodies are produced and consumed (Nast and Pile, 1998), and it is when bodies are seen as 'out of place' that the workings of power in place may be most obvious. Cahill and Riley (2001), for example, note how the privatization of aspects of femininity in Western societies means women are more vulnerable than men to the 'deviancy' label attached to visible body art, limiting the use of body art as a way of appearing attractive and as a means of identification without reference to their sexuality. Consequently thinking about bodies and the spaces through which they move can provide important insights into the way 'the same bodies are regulated differently in different spaces' (Holliday and Hassard, 2001).

Discourses circulating around feminine and masculine bodies are related to the presence or absence of fat, health, fitness and sexual attractiveness (Bell and Valentine, 1997), so it is not surprising that body size and shape become markers and means of inscribing identity. Western cultural assumptions have tended to present fat bodies negatively (Gamman, 2000). Despite an increasing range of food fantasies and lifestyles, the connection between notions of overconsumption and fat bodies remains, which can result in eating (especially for women) being a source of anxiety, and particularly so when consuming in public places.

Ageing bodies, too, can be marked in particular ways through consumption, with commodities presented as a mean of retaining youth and health (Gibson, 2000). Discourses of leisure and ageing and the meanings inscribed in them by individuals may operate powerfully to constitute identities, and to locate the bodies attached to these in particular spaces. Laws suggests that 'landscapes emplace identities which are constituted in embodied forms' (1995: 254). She believes identities are created by processes of representation, imposed on people by external sources (such as ageist stereotypes) or internalized (when such representations are accepted). If these representations are self-nominated and internalized then embodiment is said to have occurred. Identities which are externally derived are emplaced in spaces external to the subject. For example, discourses of ageing which constitute ageing as dependency and withdrawal from public life may emplace the individual in private landscapes such as 'the home, the old folks' home or the retirement village'.

The case study in Box 4.3 demonstrates how negative representations of ageing bodies are both resisted and reinforced through experiences of purchasing and living in retirement village dwellings in later years of life. Ageing consumers project their sense of place and identity onto the specific space of the retirement village before and after purchase in ways which are both imagined and real.

BOX 4.3 EMBODIMENT AND EMPLACEMENT:
LANDSCAPES OF 'LATER YEARS'

Glenda Laws (1995) indicates renegotiations around age based identities have led to the formation of a new identity, 'the active retiree'. The concept of the 'active retiree' is based on conspicuous consumption of positional goods and consumption practices, particularly those associated with participation in leisure outside individual home 'space'. In the US, Del Web Corporation's 'Sun City' retirement communities have reinforced and emplaced the identity of the 'active retiree'. These retirement communities provide homes for primarily middle class, white, Protestant and wealthy residents. Sun City communities are always built around a golf course, and include swimming pools, bowling greens, fitness centres, retail outlets, worship sites, restaurants and financial and professional services. Those whom Laws (1995) believes choose to enter an aged community are making a choice about their identity and a statement about their lifestyle.

In New Zealand/Aotearoa in recent years, gated retirement communities advertising lifestyle opportunities based around consumption and leisure have also become more conspicuous (with over 200 built since 1980). While there are some similarities to the

FIGURE 4.1   Retirement villages are marketed to potential New Zealand consumers as a lifestyle choice

US situation, the socio-spatialization of the 'active retiree' is worked out differently in retirement villages in New Zealand.[3] As in the US, entry to these communities is primarily an option for the middle class, as most residents use capital from a previous housing asset with which to purchase the commodity. A textual reading of advertisements for retirement villages also revealed these communities were being marketed as a lifestyle choice, a consumption choice which offers an active lifestyle, a sense of belonging, a means of slowing, even halting the inevitable decline associated with old age (Mansvelt, 2003; see Figure 4.1). But in exploring how these spaces were purchased and experienced through focus groups with residents, differences emerged in how the consumers themselves constituted ageing bodies, and how this was articulated with advertising rhetoric.

Few residents talked about the 'purchase of a leisured lifestyle' or village membership as a positional good. Residents did not seem to embody the active retiree construct[4] and instead suggested the prospect of companionship was more significant in influencing the decision to move. Death or illness of a spouse, failing health, fear of being/feeling alone, of not being able to access assistance, of not coping with housing maintenance, were based in the corporeality of ageing bodies and cited as reasons for purchase. Purchase of and living in retirement village dwellings were seen as means of adaptation and coping rather than resistance to their embodiment as old. As one resident stated, 'The only trouble is we're all old. Everybody's old.' Living in a retirement village was for the majority not about the resort or the leisured lifestyle promoted in advertisements. The key appeared to be in the consumption choice – choosing this form of living which enabled them to remove 'the weight of signification' (Benson, 1997) from ageing bodies, to perform, transform and reappropriate the embodiment of ageing. The choice was made possible by access to economic capital but it was one which primarily rested in the autonomy of self – in the positive choice made about one's future, in the freedom to choose, and in the ability to retain one's independence and some control over the places and people one interacts with in the later years of life.

Somewhat ironically the emplacement of ageing bodies in gated communities as spaces of freedom, youth, activity and pleasure, appears to operate to perpetuate rather than negate constructions of ageing as dependency and decline. This occurs through the spatial separation of ageing bodies, and through residents' identification of 'others' – those whose bodies are not found in assisted living spaces and who are excluded from participating in such places.

The case study in Box 4.3 demonstrates the complexity of the embodiment and emplacement of ageing identities in a small number of retirement villages in New Zealand. Consumption of this type of age-exclusive dwelling shows how place is implicated in creating boundaries between self and others and how corporeal bodies matter (both literally and symbolically) to the emplacement and embodiment of (consumer) identities. Geographers have demonstrated that places are critical in the formation of identities, providing a repository of social meanings. Presences and absences in the built landscape (such as lack of commercial spaces for different social groups) can also lead to particular consumption practices and identifications (Mort, 1998). The next section examines the concept of performativity as a means of comprehending how identities are inscribed in bodies, how agency and subjectivity operate through social roles and the practices which both make and 'locate' consumption.

Linking Embodiment and Emplacement

Performance

The case study of ageing New Zealanders examined the notion of embodiment, but the concept was defined narrowly through the acceptance of emplaced representations, with the implication that identities are always freely and consciously chosen. However, 'acceptance' is in part produced by 'performance', with embodiment comprising a process constituted via a complex network of material and discursive practices (McCormack, 1999). The research of Erving Goffman was instrumental in highlighting how such practices were linked to spatial and social settings (see Box 4.4).

BOX 4.4 GOFFMAN: PERFORMANCE IN FRONTSTAGE AND BACKSTAGE SETTINGS

Erving Goffman's (1971 [1959]) work on performance is useful for embarking on an exploration of how identities may be embodied and emplaced. His dramaturgical approach links processes of body management and identity formation with social practices that occur in everyday settings. Goffman examined matters of self-production in which interaction with others was viewed as a performance which involved the presentation of self. Identities were therefore performed, with the social roles portrayed being influenced by the audience and the social setting in which the performance occurred.

The concept of frontstage and backstage settings is important to Goffman's ideas of how the self is presented. The public performance which emerges as a self-managed series of façades occurs for the audience at the frontstage. As a collective representation the front establishes the proper setting, appearance and manner for the social roles and relationships assumed by the actor. The backstage represents the area where a different form of presentation (and by implication a more truthful one) occurs, one in which preparation for the performance takes place or where the impression fostered by the presentation is knowingly contradicted or concealed, though it may still occur in relation to a presumed audience. Goffman's theory has contributed to understandings of the actions of retailers and operators of spaces in which service and presentation of 'self' become part of exchange and the 'public' consumption experience (as in fast food restaurants, retail outlets, theme parks, banks and so on).

Crang (1994) utilized Goffman's ideas in his study of the backroom and frontroom behaviour of management and staff in a restaurant setting. McDowell and Court (1994) have also employed Goffman's ideas in the finance sector, suggesting that everyday work involves the construction of a gender performance in which 'authentic' presentation of sexual identity is an integral part of selling a particular product. Taking Goffman's work further, Hochschild (2003) conducted research with flight attendants and bill (debt) collectors to explore the commercialization of human feelings. Hochschild noted that the requirements of 'emotional labour' – involving the production and suppression of feelings of love, envy and anger in the form of appropriate facial and bodily displays in public or 'frontstage roles' – could actually impair the emotional functions of individuals, estranging them from aspects of the self.

However, Gregson and Rose (2000) suggest there is a number of difficulties with the implementation of Goffman's ideas in geography. First, Goffman's analysis presumes an active, prior, conscious, intentional and performing self (2000: 433), an essential and authentic self that exists always outside the performance. Second, the notion of performance positions subjects in relation to a reactive, interpretive 'audience', with the consequence that the audience's agency and power as performing subjects are removed. Third, the social context becomes a pre-existing and empty stage which is 'mapped' through performance rather than being enmeshed in the constitution of the performance. Consequently the interrelationality of subjects is removed and power exists apart from the performer (2000: 445).

Performativity

Gregson and Rose's critique of Goffman's work touches upon the issue of power. How identities are performed is connected to one's location in the social world and to structures and memberships of social groups, classes and communities (Entwistle, 2000). Performance is also connected to the technological, physical, social and economic environments in which the body has its being (Harvey, 1998) and is not independent of discourses against which our evaluation of our own bodies proceeds (Valentine, 1999b). Judith Butler, in considering the power effected via discourses of gender, suggests 'performativity' is a more appropriate concept with which to grasp the relationship between power, practice and identification (Box 4.5).

BOX 4.5 BUTLER: PERFORMATIVITY, GENDER AND IDENTITY

Geographers Nicky Gregson and Gillian Rose (2000) argue that performance is subsumed within and always connected to performativity. In doing so, they draw on Judith Butler's work on gender and identity. Butler (1990) argued gender is a performance; it

> is what subjects do, say and act, rather than what they are. 'Gender is always a doing, though not a doing by a subject who might be said to pre-exist the deed – there is no gender identity behind the expressions of gender; that identity is performatively constituted by the very "expressions" that are said to be its results' (1990: 25). Thus, for Butler, identities do not pre-exist performances and are not freely chosen.
>
> Butler's concept of performativity encompasses notions of discursive power which operate *through* the performance to constrain and enable subjects and performances. Performativity comprises the citational practices which reproduce and/or subvert discourse (Gregson and Rose, 2000). The repetition of acts creates norms, which are in turn maintained through their repetition. Identities are performatively enacted significations constituted through 'stylized and regulated repetitions of acts through time' (Butler, 1990: 141). Butler suggests that 'there is no possibility of agency or reality outside the discoursive practices that give those terms the intelligibility that they have' (1990: 148). However, while individuals do not have agency outside the systems of signification in which they are enmeshed, a form of agency may still be possible, centering upon how to repeat acts, and whether to repeat and displace the norms that enable the citational practices to occur (Butler, 1990).

Gregson and Rose (2000) develop Butler's ideas by considering how space is a performative articulation of power. The empirical becomes the site through which and in which (powerful) 'citational practices are displayed, re-enacted, resisted and transgressed' (2000: 435). Specific performances do not occur on pre-existing 'stages', rather they bring these spaces into being (2000: 441). For example, car boot sales are not empty 'stages' but are events which depend on their 'being' through specific performances by promoters, marshals, vendors and buyers with the blurring of boundaries between audience and performer (2000: 444). Butler's concept of performativity can be used to understand how these performers are saturated with power and engage in highly gendered practices of buying and selling, which in turn re-cite (and reinscribe) heterosexual norms and dominant understandings of gender. Car boot sale performances can also be viewed as existing in interrelationship with other performances, audiences and powers, exposing the power relations that work in conventional retailing (Gregson and Rose, 2000).

Thrift (2000b; 2000d), in critiquing the emphasis on the visual and representation in geography, uses the concept of performativity to embrace the bodily practices, movements, senses and habits (which are not necessarily subject to discourse) by which human beings meaningfully engage with and transform the world. His work appears to counter some of the concerns of Nelson (1999) that performativity reduces subjects to compelled, unreflexive performers of dominant discourses and that thereby we 'miss the how and why of human subjects doing identity, a process directly tied to their lived personal history, intersubjective relationships and embeddedness in particular moments and places' (1999: 349). Thrift (2000c) explores 'performing cultures', demonstrating how new 'fast' subject positions in business management create new performative spaces, arguing social practices have citational force because the space in which they are embedded makes room for non-representational qualities (such as serious play and self based techniques to cope with management dilemmas).

Thus if body practices and performances are incorporated within notions of performativity the concept provides a valuable means of understanding how discourses, practices, bodily actions and surfaces together create geographies of consumption – geographies in which subjects are constituted, embodied and emplaced. Louise Crewe, Nicky Gregson and Kate Brooks' research draws on these ideas to show how rituals of exchange, possession, appropriation and disinvestment are performative of bodily discourses and subjectivities (Box 4.6).

BOX 4.6 BODY MATTER: DIRT, DISCOURSE AND SECOND-HAND CLOTHING

Crewe, Gregson and Brooks' research (Gregson et al., 2000; 2001a; 2002b) involved participant observation in shops and interviews of employees, consumers, managers and owners in charity and retro clothing shops. They found that discourses of the body framed the sale and purchase of second-hand clothing. Retailers and voluntary workers were involved in expunging bodily traces from previously worn garments through sorting donated clothes and purifying rituals such as washing and ironing (Gregson et al., 2000). Charity store workers valued clothes which held fewer traces of previous bodies more highly and clothes that were disposed of usually bore the personal imprint (literal or metaphoric) of owners, for example underwear, shoes, night attire, filthy and/or worn out items. Yet value itself is not stripped away by removing traces of previous occupiers, it is 're-contextualized' – for many shoppers the value of second-hand clothing is actually derived from the fact that this clothing had been previously worn (Miller, 2000: 80).

From the potential consumer's point of view, body matters were significant too. The risk associated with buying second-hand clothing was about engaging with unfamiliar others, and unknown commodity biographies. Clothes that were hand-me-downs (Gregson et al., 2000: 109) and children's 'pre-loved clothing' (see Clarke, 2000) have quite different bodily inflections, valued in part because of the presence rather than absence of previous owners (see the case study on transnational biographies in Box 5.4). Clothes 'closest in' were normally taken home and washed, a process of both disinvestment (of the old body) and re-enchantment (of the new body to which they now belonged), and the more serious second-hand shoppers often chose the more risky course of buying 'closer in'.

Body matters were also 'place' matters. Drawing on Goffman's ideas of frontstage and backstage, Gregson et al. (2000) demonstrated how the front zones of charity shops were associated with professionalization (the practice and performance of 'retail') and the presentation of goods as if new. Back rooms were odorous and dirty zones in which the body was omnipresent, where work was performed to eradicate bodily traces.

> In contrast, in 'retro shops' (shops that sell clothing from the recently worn past) Gregson et al. (2001a) found that apparel was not constructed so negatively in relation to body, perhaps because of greater temporal distance between consumers and previous owners. Miller (2000) suggests value here may be more a function of (past) authenticity. Cleaning rituals for retro clothes become a means of re-enchantment in a different way to that of charity shop consumers, becoming a process of reasserting links to past lives rather than expunging evidence of previous occupiers.
>
> Gregson, Brooks and Crewe's research on charity and retro shop clothing provides valuable insights into subjectivities and identification of self and other. Consuming clothing becomes an extension of corporeality (Gregson et al., 2000). The practices of charity and retro shop employees and consumers are not simply performances, they are also seen as performative – citing and re-citing discourse of the sealed and bounded (Western) masculine body, a body which is threatened by the leaky and polluting excesses (the dirt, excretions and mess) of the feminized other (see Longhurst, 2001).

Gregson, Crewe and Brooks' research on retro shops extended notions of how rituals of consumption (such as personalization, possession and transformation: Gregson and Crewe, 1997b) may be performative of particular narratives of the body and of space. In interviews with shop managers and consumers, Gregson et al. (2001a) identified two modes of valorizing retro clothing. One mode drew on the notion of 'the carnivalesque' (Bakhtin, 1984), invoking notions of fun, wild play and spectacle through collective participation (see Figure 4.2). This involved consumers purchasing and wearing retro clothing because of a desire to produce excesses of bad taste based, for example, on stereotypical shapes, styles and fashions of the 1970s. The carnivalesque was thus a transient experience based on an 'other' formed in relation to a serious stylish attired 'normal' self (Gregson et al., 2001a: 9–12).

The second mode of appreciation involved 'knowingness' – discerning authentic 1970s gear and reappropriating it within contemporary fashion (see Figure 4.3). This involved wearing garments as routine apparel rather than 'temporary or playful costume'. Here knowledge and discernment are performative, citing notions of imagined authenticity and individuality (2001a: 12–18). The complexity of such modes is revealed in their relationality; citations of bad taste and good taste mark both modes. In the 'carnivalesque' this works by celebrating and mocking bad taste as other than self, and in the 'knowingness' mode it is represented by the bad taste of unknowing others (2001a: 18). Thus research on second-hand clothing demonstrates not only how consumption acts cite powerful bodily norms but also how discourses and bodily experiences (e.g. sights and smells) surrounding 'body matter' can play a significant role in the valuation of commodities, and the embodiment and emplacement of consuming subjects.

FIGURE 4.2 Second-hand clothing in carnivalesque mode: friends dressed for a 'bad taste' night out

Placing Consuming Identities: Geographies of Food

As stated earlier, an emphasis on purchase of commodities renders invisible the multiplicity of ways in which consumers take and make meaning from consumption. In this final section of the chapter I examine how consumption of food influences the spatiality of the body. Geographies of food have provided a critical conduit through which issues of displacement, sociality, embodiment and emplacement and their construction in place might be examined.

'Eating is a useful medium to explore the space of the body because it is one of the ways that the spatiality of our bodies is brought into being' (Valentine, 1999b: 331). If 'we are what we eat', then 'we are also what we do not eat' and we are also 'where we eat' (Bell and Valentine, 1997). Food has a special status

FIGURE 4.3   Second-hand clothing worn as 'knowingness': Travis and Kelly are serious retro wearers

as a consumption good; while it is necessary to sustain life, it is also symbolic of luxury and lack. Food may be consumed in a multiplicity of ways for many reasons: survival, pleasure, anxiety or boredom (Mintz, 1993). The purchase, preparation, display and devouring of food can be a measure of economic and cultural capital and a potential source of social exclusion. Food also becomes 'the body' and waste as it is consumed (Yasmeen, 1995). As mentioned earlier, consumption of food plays a role in the differentiation of bodies and identities: the fat, the anorexic, the slob, the glutton, the supermodel are constituted around bodily practices of eating which are variously performative of discourses of desire, aesthetics, sexuality, pleasure, revulsion and self-control. Food moralities are also spatialized, for example in conventions around eating on the street, at work or at school, in public or in private.

Purchasing and eating food thus contribute to the construction, representation and spatiality of corporeal subjects and the construction of 'social imaginaries, which position individual dietary practices within wider discursive framings' (Cook et al., 1999: 223; see also Domosh, 2003). Food consumption practices are important to the construction of individual, group and place identities at scales from the body to the global (Bell and Valentine, 1997). As commodities, food (and beverages) can also be significant in working place and product meaning, as in the *appellation d'origine contrôlleé* labelling of wine (Moran, 1993). In making places as symbolic constructs, food is 'deployed in the discursive construction of various imaginative geographies' (Cook and Crang, 1996: 140; see Box 4.7).

BOX 4.7 THE WORLD ON A PLATE: DISPLACEMENT AND IDENTITIES

Cook and Crang's (1996) study of the fashioning of London as a cosmopolitan food space examined how flows of foods, people and culinary knowledge were articulated. Cook and Crang found that when people ate 'foreign' food the world on their plate was as much imagined as it was real. Geographical knowledges about where foods come from, and the objects and acts associated with them, become a means to 're-enchant' commodities and to differentiate them from the devalued functionality and homogeneity of standardized products, tastes and places (1996: 132). This occurs through processes of displacement where food consumption occurs in local contexts but those contexts are 'opened up by and constituted through and by connection into any number of networks' (1996: 138). Displacement implies that consumer identities fragment into juxtaposed 'identifications' that stretch beyond particular institutions and spaces of subjectification (Crang, 1996: 64). Consequently consumers are entangled within the geographies of commodities they consume and the contexts in which they consume them (1996: 65).

How displacement operates through the imagined geographies of self and other is exemplified in Jon May's work. Analysing consumers' talk, May (1996) demonstrated that the consumption of exotic foods by young British consumers living in urban areas rested upon the racist understandings that frame imaginative geographies of the 'other'. Exotic consumption practices in turn constructed racialized notions of 'development'. May's research shows how meanings of commodities are also transnational, with the politics of identity and place bound up in transnational commodity flows. Consumers actively construct meanings of self, others and transnational commodities in different spaces, which are themselves 'caught up' in transnational networks (Dwyer and Jackson, 2003).

Thus foodscapes – a view of place in which food is used to filter out and bring into focus certain human relations (Yasmeen, 1995: 2) – have much to contribute to understandings of consumption. Though purchase, preparation and consumption of food are seemingly banal acts, geographer Gill Valentine has demonstrated that food practices are linked powerfully to the embodiment and emplacement of identity, to the politics and sociality of relationships, and to the spaces in which such practices and meanings are interpreted (see Box 4.8 and Figure 4.4).

FIGURE 4.4   A children's birthday party. Food practices, politics and places can play an important role in shaping socialities and subjectivities, here reproducing discourses of identity, friendship and family

BOX 4.8 CONSUMING FOOD: SPACES, PRACTICES, IDENTITIES

Gill Valentine's (1999a; 1999c; 2002) research on food consumption practices involved qualitative research with households (nuclear families, young couples, retirees, lone parent families, lesbians and widowers) and institutions (including a male prison, two schools and a homeless support group). Valentine was able to interrogate participants' stories to explore how people, their experiences and their food practices were constituted and located within a repertoire of narratives. Such narratives were not necessarily of individuals' own making and were often located within 'wider multiple plots of family, work institutions, nation and so on' (1999c: 496).

Valentine noted how food 'becomes the corporeal body' – how as a tactile space it is always sensing and engaging with itself (how we feel inside) and the world (outside) to produce its spatialities (1999b: 331). The narratives through which these bodily experiences were expressed were complex. Consuming food could bring ambivalent and even contradictory experiences, for example of pleasure and pain, freedom and guilt, and it made individuals in her study feel variously happy, guilty, sexy and sluggish.

Valentine (1999b) found there was considerable pressure for women to produce the space of their bodies in a sexually desirable way. While many women in Valentine's (1999a) study did judge themselves on how they 'looked' to themselves and others, they also judged their attractiveness on how they felt about their bodies. For many women narratives of identity could be understood as 'bodily inflections of wider socio-spatial sexual relationships' (1999b: 333). Consequently women in the study talked about disciplining bodies through diets or exercise – to make themselves more desirable. For the men in the study the picture was somewhat different. Despite the increasing commodification and aestheticization of men's bodies in the media, Valentine's study found that appearance or attractiveness was not a significant feature of men's narratives; their bodily experiences were constructed via discourses of health and functionality, rather than sexuality as was the case for women.

Social practices should not be seen as determining factors in the constitution of identity, as the social, spatial and political context in which identities are formulated is also important. Valentine illustrates this point with regard to 'Carol', who continues to identify as a meat eater despite the fact that she is cooking and eating vegetarian meals for herself and the members of her family (1999c: 500). Carol also demonstrates both the commensality and negotiation of eating practices (one of the children was the first to become vegetarian and other family members eventually began eating vegetarian meals) and how differences in identities can be reconciled to produce uniform consumption practices (1999c: 500).

Thus the meanings given to food and food consumption practices, and the politics of household consumption, are all embedded in complex and shifting narratives of identity which may be both regulatory and enabling. Anna's food biography demonstrated how wider changes in narratives of identity are relational (Valentine, 1999c). Anna's subjectivities as a student, an accountant and an environmental worker were 'articulated on her plate at home', influencing what food she bought to eat (and where from) and how she created narratives of self which in turn influenced her politics, her employment and her social networks' (1999c: 515).

Gill Valentine's work has revealed the ways in which consumption is corporeal, and is implicated in narratives of identity which extend beyond (and make fluid) spatial boundaries of self, home and work. Her research demonstrated how food consumption is significant in the constitution and expression of these narratives – narratives which are constituted in 'moral' discourses which are sexualized, gendered and raced and which are performative of 'good parenting', 'desirability', 'healthy and/or fit bodies' and 'proper and civilized' food practices and habits (Valentine, 1999b).

Valentine's study of household practices also illustrates the politics of household consumption, a politics which may be based around individuality, sociality, sharing, negotiation and conflict. Ultimately she illustrates that food politics and practices are not only identity issues, but body and space matters too.

Commodity and Body Journeys

Miller (1997: 45) argues that consumption not only expresses relationships, but also becomes a primary form through which our very understanding of what it means 'to be' is expressed. The geographical work in this chapter provides an important critique of narratives of consumption which present consuming as a (superficial) enterprise based on the playful/anxious search for identity through the purchase of commodities. If the body is the key location of social experience then the spaces through which it moves and is disciplined, displayed, performed, felt and experienced really matter (McDowell, 1995: 76).

The question of how the cultural and the material are infused to create identity would seem a significant one. Harvey argues: 'The study of the body has to be grounded in an understanding of real spatial–temporal relations between material practices, representations, imaginaries, institutions, social relations, and the prevailing structures of political-economic power' (1998: 420). Thus it is important to understand how places are not simply settings, but are constitutive of individual and collective consumption practices. It is also essential to explore how particular subjects are embodied and emplaced as consumers (such as 'the rich', 'the poor', young, old, gay, health conscious or ethical consumers), and to examine how material and bodily consumption practices shape social relationships and identifications of self and others. Concepts of performativity and displacement have provided insights into such issues, emphasizing the connections between material and discursive practices in place, and across space.

The geographical research presented in this chapter has also examined the ways in which particular commodities may operate as media for the expression and creation of identities. Research on a variety of 'positional' and mundane goods and services as part of material culture is also essential to explore both the social lives of commodities in place, and the 'placed' role of commodities in social life. While there is a substantial amount of literature on how commodities are given meaning and incorporated into everyday life through rituals of performance and possession (Gregson et al., 2000), much less has been done on how they are decommoditized, devalued, recycled or end up as trash (though see Hetherington, 2004).

In highlighting the complex relationship between consumption practices, commodities and place, geographers have argued that commodity purchase, use, meaning and experience cannot be reduced to 'identity value'. However, consumption practices are complicit in the creation of boundaries between self and others through processes of embodiment and emplacement which are spatially, socially, morally and politically constituted. The bodies, identifications and practices of consuming subjects are not trivial, but are vital to how people connect with and construct 'the world'.

FURTHER READING

Bell, D. and Valentine, G. (1997) *Consuming Geographies: We Are Where We Eat.* London: Routledge.

Cook, I. and Crang, P. (1996) 'The world on a plate: culinary culture, displacement and geographical knowledges', *Journal of Material Culture*, 1 (2): 131–53.

Gregson, N. and Rose, G. (2000) 'Taking Butler elsewhere: performativities, spatialities and subjectivities', *Environment and Planning D: Society and Space*, 18 (4): 433–52.

Gregson, N., Brooks, K. and Crewe, L. (2000) 'Narratives of consumption and the body in the space of the charity shop', in P. Jackson, M. Lowe, D. Miller and F. Mort (eds), *Commercial Cultures: Economies, Practices, Spaces.* Oxford: Berg. pp. 101–21.

Longhurst, R. (1997) '(Dis)embodied geographies', *Progress in Human Geography*, 21 (4): 486–501.

Valentine, G. (1999) 'Eating in: home, consumption and identity', *The Sociological Review*, 47 (3): 491–524.

Valentine, G. (2002) 'In-corporations: food, bodies and organizations', *Body & Society*, 8 (2): 1–20.

NOTES

1. I refer to identity as one's subjective sense(s) of self. There are numerous approaches to how identity formation operates, and in this chapter I mainly draw on the 'subject of language' approach. This work suggests identity formation is constituted discursively in and through the production of difference between self and others. Identity formation is also related to a politics of location, the interdependency of the material and the metaphorical, and the relations of power as they are grounded in socio-spatial contexts.
2. Callard (1998: 392) argues that the assumed connection between purchase of commodities and identity as part of a postmodern epoch means terms such as 'fluidity', 'fragmentation', and 'hybridity' lose their subtlety and their grounding in the specific meaningful theoretical and empirical contexts in which they arose.
3. For the purposes of this research, a retirement village is a resident funded complex where mature persons (normally age 55+) purchase or acquire the right to accommodation and services, and often the use of shared 'community' facilities. Services paid for by residents may include local authority, sewerage and water rates, fire insurance, maintenance of properties and gardens, emergency call systems, meals and nursing assistance.
4. The active retiree lifestyle appears to be constituted quite differently in New Zealand, with older people feeling guilty for engaging in leisure activities which have no extrinsic or productive outcomes other than self-pleasuring (Mansvelt, 1997).

5

Connections

One enduring dilemma in the consumption literature is how to make sense of the interdependencies of consumption and production without positioning the two spheres as dichotomous categories, or reinforcing these by assuming that consumption is a cultural phenomenon and production an economic one. This chapter covers various perspectives which address the connections between production and consumption. Viewing production and consumption in particular ways has consequences for how power and agency are constituted. Differing modes of connection are used to challenge the notion of consumption as a bounded sphere and the essentialist constructions of consumers and consumer spaces which derive from this. There are three modes of connection which I explore in this chapter: chains, circuits and actor networks.

Linking Production and Consumption: the Commodity Chain

The metaphor of the chain which connects production and consumption processes is a powerful one, imposing a structure on social relations across particular places. Commodity chains comprise 'links' representing 'discrete, though interrelated, activities involved in the production and distribution of goods and services' (Blair and Gereffi, 2001: 1888). The concept of a commodity chain has been used by geographers as a means of studying the movement of a commodity from conception to disposal through various nodes or phases such as agricultural production, processing, manufacturing, distribution, marketing, retailing, purchase and use. What has been of particular significance to geographers is the 'placing' of these nodes and the social and spatial connections between them (see Box 5.1).

BOX 5.1 CHICKENS AND COMMODITY CONNECTIONS

Michael Watts describes how he has brought an oven-ready chicken into his lectures and asked his students to identify it. He tells his students that what they are actually looking at is 'a bundle of social relations' (1999: 307). Watts suggests that in

> America consuming a chicken connects consumers to a wide range of people, places and processes. This includes farmers as battery producers, feed companies, and scientific research and development organizations which promote genetic control, reproduction, disease and growth control in the birds, so that the chicken becomes an industrialized commodity which is part nature, part machine. Consuming chicken also connects one to transnational corporations who produce the chicks and food, contract farmers who grow chickens, and the poultry processing industry. The processing industry in the USA is 'one of the most underpaid and dangerous in the country' and employs a significant proportion of immigrant labour (1999: 307). Watts demonstrates how chicken is a global commodity, with the USA being the world's largest producer and exporter of broiler chickens in a market which is highly segmented geographically (e.g. American consumers prefer breast meat, while legs, feet and wings are the preference of Asian importers). Not only does this example demonstrate the complexity of the production system for the chicken, but in Watts' words 'You start with a trivial thing – the chicken as a commodity for sale – and you end up with a history of post-war American capitalism' (1999: 308).
>
> Watts' illustration of how consuming chicken connects his students with a multitude of processes and practices is a powerful one. It demonstrates how studying the biographies, the geographies of commodities and places, and the processes through which they are transformed presents possibilities for thinking about the structure, organization and governance of the connections between production and consumption and how these are constituted in place and across space. This is a theme which has been taken up by geographers exploring different modes of connection.

Underpinning much commodity chain literature is a vertical approach to studying consumption. This involves exploring linear changes in commodity processes, relationships and practices across various sites. This approach aims to uncover the social relationships behind the production of a particular commodity. Applying the chain metaphor to production and consumption relationships enables one to explore commonalities and differences in commodity trajectories and has the potential to illuminate 'some of the key characteristics of contemporary capitalism, and the dynamics of change which have emerged in the age of globalization' (Raikes et al., 2000: 409).

Two intellectual traditions have been encapsulated in the chain metaphor: the global commodity chain literature which focuses on the organization and control of chains across space, and the systems of provision literature which follows commodities along chains.

Global commodity chains

Initial research on commodity chains tended to focus on the dynamics of the connections between production and consumption manifest across national borders. Global commodity chain literature is derived from world systems theory[1] and has emphasized political-economic systems of production and consumption rather than individual nodes or the particularities of place (Leslie and Reimer, 1999). Much of the early global commodity chain (GCC) literature focused on analysis of industrial commodities, depicting how commodities were produced in peripheral regions of the world for consumption by a core of countries (see Gereffi and Korzeniewicz, 1994).

In documenting how commodities are made and move across space, the GCC literature has highlighted the spatiality of production/consumption processes and the role of powerful agents in transforming and regulating production and consumption. 'Agents' such as manufacturers, buyers and distributors play an important role in mediating the social relationships and flows of materials, peoples and knowledges associated with them. Their actions can be interpreted and situated in social-spatial settings, institutional contexts that are constituted by particular political, economic and social relations. The global commodity chain approach has consequently been invaluable in reflecting on the form and nature of place based differences in which chains are formed and operate, and the outcomes of particular linkages across space. Global commodity chains are strongly associated with economic aspects of globalization, the formation of spatial divisions of labour, and the social and spatial inequalities arising from this (Blair and Gereffi, 2001).

Firms are critical links in the chain, with power vested in lead firms which drive the chain by coordinating and controlling the organization of production. Gereffi (1999) argues the type of lead firm will determine the type of governance which characterizes a chain, and as a consequence the contours of local development where chains 'touch down'. The GCC approach stresses the organizational and institutional aspects of consumption and production, especially with regard to how key agents set up and maintain production and trade networks (Raikes et al., 2000: 394). It is argued there are two modes of organization governing commodity chains: buyer and producer driven forms (see Figure 5.1).

Producer driven chains are characterized by control over the chain by vertically integrated transnational corporations producing commodities. The growth of these firms is linked to the (theorized) development of a Fordist regime of capitalist accumulation after the Second World War. Import substitution policies of developed and developing nations and export processing zones also facilitated the establishment of producer driven chains (Gereffi, 2001). Producer led chains (or supplier driven chains as they are also called) are typified by capital and/or technology intensive industries such as automobiles, aircraft and computer firms. In contrast, in buyer driven chains, the form and operation of the chain are directed by firms involved at the consumption end (such as retailers, marketing and design companies) who dictate manufacturing and/or production processes. These firms procure commodities according to their specifications, often dictating the form and organization of production and supply (see the Nike case study in Box 5.2).

Gereffi (2001) argues that the emergence of buyer driven chains in the 1960s was assisted by advances in transportation and communication and a shift in the

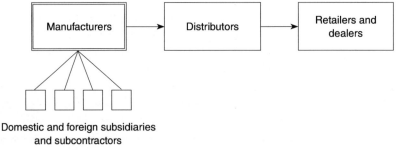

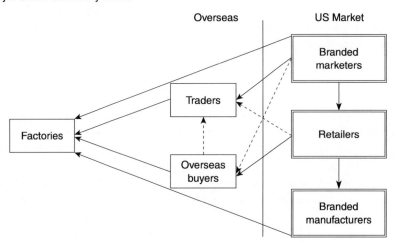

FIGURE 5.1 The organization of producer and buyer led commodity chains (Gereffi, 1999)

industrial strategies of developing countries from import substitution to export oriented growth (encouraged by US government, International Monetary Fund and World Bank policies). Increasing competition between developed country retailers and markets compelled firms to become involved in the growth and organization of offshore sourcing networks (2001: 32–3). Buyer led chains have been associated with labour intensive industries such as apparel, footwear, toys and consumer electronics, and Gereffi argues their formation is part of a general transformation from 'manufacturer shift' to 'consumer pull'. Many of these firms are brand led (where the identity or image rather than the characteristics *per se* of the product or service is predominant in its marketing). A consequence of the emergence of buyer driven chains has been the relocation of the most expensive aspects of the production process to those areas with abundant and low cost labour. This has often resulted in a geographical separation of production processes, with control of flows of information, labour, skills, products, logistics, marketing and design retained in core countries, and manufacturing, processing or assembly located in 'peripheral' countries with lower wage and unionization levels and where employee rights and worker insurances are less well guaranteed. Unlike producer driven chains which are often vertically integrated, buyer driven chains can consist of a variety of independent enterprises which are connected through both formal and informal arrangements (such as alliances or subcontracting arrangements) (Gereffi, 2001).

The classification of producer or buyer led chain is useful for beginning to think about issues of power and agency as expressed though production–consumption flows and relationships. However, the increasing complexity and breadth of chains in a globalizing world means that discerning how relations of power and control are established across chains is challenging. The growth of Internet mediated business also complicates Gereffi's notion of buyer/supplier led chains. Gereffi (2001) suggests a third form of governance, organized around the Internet, has emerged in the mid 1990s but he is unclear as to the implications of this.

The GCC literature is linked to notions of globalization as an economic and political project managed through capitalist institutions (largely in the form of transnational corporations), financial institutions and regulatory infrastructures (Whatmore and Thorne, 1997). This has implications not only for how consumption is seen (that is, as a mask for exploitative capitalist relationships) but also for the political economy of development itself. The work has tended to emphasize the effects of production in the periphery and has presented consumption as a starting point from which to embark on a more significant exploration of production (Leslie and Reimer, 1999). Despite this, commodity chains have provided the basis for the political mobilization of individuals through consumer campaigns aimed at encouraging companies to improve their workplace practices and to secure better pay and conditions for employees. Protest has also centred on influencing state and local state legislation, and the neoliberal regulatory policies of the IMF and the WTO (Blair and Gereffi, 2001). Campaigns have generally sought transformation through increasing consumer awareness and influencing purchasing decisions (see Box 5.2 on Nike, a company whose branding strategies and workplace practices have attracted considerable academic and public attention).

**BOX 5.2 NIKE: 'JUST DO IT' OR 'JUST STOP IT'?
CONSUMER ACTIVISM AND GLOBAL CHAINS**

'Consumers and workers of the world unite – Just do it! If you do, you can affect the behaviour of manufacturing giants such as Nike, for whom image is everything.' This was the opening line of a newspaper article in the *Washington Post* on 15 May 1998 about Nike's decision to improve conditions for its workers in plants overseas (Dionne, 1998).

Nike Inc. is the world's largest athletic footwear and apparel company. The Nike brand and the 'swoosh' icon have how become, for many, instantly recognizable symbols. Such is the power of the Nike brand that Lury (1999) argues its repetition reconfigures space and time, enabling the durability of the brand to become detached from the durability of the product. As a consequence the objective properties, uses and meanings of a Nike commodity actually becomes an effect of the brand. The Nike brand combines the power and emotion of sport with innovation at the cutting edge, an image which is encapsulated in the maverick brilliance and competitiveness of the rebel. Nike's choice of high profile 'athletes with attitude' (such as John McEnroe, Andre Agassi, Michael Jordan and Kathy Freeman) has been critical in heightening brand awareness. Nike revenue totalled US$10.7 billion in 2003 (Nike Inc., 2003).

The company has its origins in 1964, when CEO Phil Knight began selling sports shoes imported from Japan from the back of his truck. Today Nike is a major design and marketing company which has promoted the business philosophy of 'no limits' spending on branding (Klein, 2000). Though Nike Inc. is not directly involved in the manufacturing of its products, the firm manages to control how, where and when its consumer goods are produced through subcontracting out its manufacturing to independently owned and operated factories (Korzeniewicz, 1994). A complex three-tiered system of contracting enables Nike to shift its resources and production geographically in response to changes in the factors of production and consumer demand (Donaghu and Barff, 1990). Such is the complexity of Nike's subcontracting, supply and financial arrangements (see Goldman and Papson, 1998) that the metaphor of a buyer led chain sits uneasily (Jackson, 2002b). Nike related factories are thought to employ around 500,000 workers (Larimer, 1998), with the majority of Nike shoes and much of their clothing being produced in factories in Asia (Schoenberger, 1998). Production occurs in countries such as Thailand, Vietnam, Indonesia and China where labour costs are low and possibilities for unionization may be limited. A pair of Nike sports shoes which costs around US$16.70 to manufacture sells for approximately US$100 around the US (Beder, 2002).

Throughout the 1990s a series of campaigns has been running which intend to alert consumers to various human rights abuses involved in the Nike 'commodity chain'. These human rights issues include accusations of physical, mental and sexual abuse in factories, poor and unhealthy working conditions, and the lack of opportunity to form unions (Connor and Atkinson, 1996). Controversy has also centred on the payment of workers at rates less than the minimum wage or, where minimum wages have been paid, on the concept of a just or fair 'living' wage (see Schoenberger, 1998). While a few campaigns have encouraged consumers to boycott Nike products and sponsorship, many others see this as harming workers in the countries in which shoes and apparel are produced (Oxfam Community Aid Abroad, 2003). Protest rhetoric has often focused on discourses of 'morality', and the relative inequality between wages and the spending on advertising and promotion (with frequent quotation of the estimated $20 million per year golfer Tiger Woods is paid to endorse Nike products). The Internet appears to be a major means of disseminating information about the company and protest activity.

A search on the web reveals 'Nike – Don't Do It', 'Nike – Do It Justice', 'Just Don't Nike', 'Just Boycott It', 'Ban the Swoosh', and 'The Swooshtika' as slogans which contest Nike's brand image and which have been the basis for consumer campaigns.

Hartwick (1998) suggests the dependence on image makes corporations such as Nike Inc. vulnerable to boycott and to consumer pressures which include semiotic contestations of sign meanings. By 1997 the company had become 'a symbol for sweatshop labour in the third world' (Beder, 2002). Anti-Nike campaigns have sought to promote consumer awareness of the conditions in which Nike products are produced, and a consideration of issues of social justice.[2] Anti-Nike campaigns have endeavoured to mobilize individuals to become active in organizing and participating in protests, lobbying organizations (such as schools and universities whose sports teams wear their products)[3] and sending letters, faxes and e-mails to Nike Inc. The sovereignty of the 'knowledgeable' consumer is often stressed, with individuals encouraged to protest outside Nike retail outlets and hand out leaflets informing shoppers about workplace practices. Johns and Vural note how the 'Stop Sweatshops Campaign' was an attempt to mobilize not just union members but a 'community of consumers' (2000: 1195). Targeting retailers rather than focusing on production processes, the campaign was concerned with encouraging consumers to pressure retailers to eliminate oppression down supply chains and to make manufacturers and retailers responsible for the commodities they make and sell (see Figure 5.2). This campaign was instrumental in the

"Raw materials from third world, manufactured in second world, marked up in first world."

FIGURE 5.2   The politics of commodity chains made visible (cartoon by Ted Goff)

> passage of the Stop Sweatshops Act in 1997 in the USA, in lobbying the President's Apparel Industry Task Force, and in the Sweatshop-Free Cities movement. Nike and Liz Claiborne were two of the firms who agreed to comply with a workplace code of conduct that established standards for decent working conditions as part of the 'No Sweat' Agreement of 1997 (Hartwick, 1998).
>
> Until the late 1990s Nike company representatives operated a discourse of denial, consistently dismissing the claims of consumer activists and claiming the company did not 'manufacture' sports shoes. However, in May 1998 Phil Knight responded publicly to criticisms and announced a series of measures aimed at improving conditions of work in its factories. Though still refuting reports of abusive labour conditions, Knight noted how the unfavourable publicity had damaged his company's image. He subsequently announced plans to revise the firm's code of conduct (first formulated in 1992) to ensure Asian factories complied with US workplace air quality standards, to increase the independent monitoring of factories, to ban children under 15, and to pay the local minimum wage. The anti-sweatshop campaign has been credited with leveraging change in company practices, but it is impossible to know the extent to which the unveiling of the commodity fetish (through publicity about workplace practices) translated to consumer activism or to action on the part of the company to change its policies and practices. Discourses of transparency, social responsibility, public relations and consumer agency may be framed, imagined, interpreted and practised quite differently by corporations and groups of consumers (see Silvey, 2002).

Hartwick (1998) argues a new kind of politics is emerging, one which links consuming and consumers' institutions with institutional actors upstream. Making consumers aware of production practices as an underpinning of activist strategies rests on a conception of the naive consumer and may privilege academic understandings of commodity flows (Jackson, 1999). Nevertheless, the emergence of new forms of industry governance (such as the 1997 'No Sweat' Agreement) does accentuate the agency vested in consumers (Johns and Vural, 2000). If a new politics is emerging, it cannot be considered unified but instead should be seen as a consequence of the construction of situated communities of interest which are fractured and diverse (Johns and Vural, 2000)[4].

Much of the work on global commodity chains has centred on the 'production' side of commodity chains and revolved around the analysis of industrial commodities and agro-commodity chains. It has also been applied to tourism, services, fresh fruit and vegetables, footwear, electronics and automobiles (see Raikes et al., 2000). Consumer services such as tourism, healthcare, financial services, accommodation and travel services are less well examined and less easily applied to the concept (Clancy, 1998: see Box 5.3). During the last decade retail geographers have devoted a considerable amount of attention to the capacity of buyer led firms to influence production, sales and retail spaces. Global commodity chain literature has contributed to identifying shifts in relationships between manufacturers and retailers. Organizational and technological regulatory changes (changing firm relationships) have highlighted changing forms and operations of institutional power. For example, Burt and Sparks' (2001) work on retailer Wal-Mart demonstrated how power is affected through the chain when suppliers do not give up ownership of their goods until they are sold to the customers. This has the effect not only of improving Wal-Mart's cash flow, but of making suppliers much more sensitive to consumers, stocking levels and supply systems.

BOX 5.3 SERVICES AND COMMODITY CHAINS: THE CASE OF SEX TOURISM

Clancy has endeavoured to apply the global commodity chain approach to his research on sex tourism. He suggests sex tourism has become the ultimate form of 'modern' tourist consumerism, where the 'sexual conquest of host women, men or children' becomes another tourist souvenir (2002: 73). Though the organizational or governance structures of the chain do not easily conform to buyer or producer driven models, with production and consumption of the sex service often taking place in the same locale, the application of a GCC approach to sex tourism is useful for highlighting power and exchange relationships (Clancy, 1998: 129). His work has demonstrated how the bulk of sex tourism is North–South in nature, often drawing on representations of (primarily women) in the global South as primitive, exotic, uninhibited. He explores why the sex tourism commodity chain touches down in Cuba, linking this to economic crisis and its effects on women, its lower relative cost in a price sensitive industry and government approval through acquiescence. Clancy notes how the sex services chain mirrors the power relations in other buyer driven chains. Brokers (as pimps, bar owners, tourist operators and so on) control the production process and extract surplus value far in excess of the 'producers' of such services. Clancy's study demonstrates some of the insights of GCC literature, illuminating links between production and consumption at the local level and making sense of this in relation to other scales.

Though Clancy's (2002) research touches upon some of the broader social relations which constitute the consumption, representation and reproduction of sex tourism, the intersection of this chain with other horizontal processes – such as how sex tourism is materially and symbolically (re)produced, mediated through the Internet, marketed and advertised – are not examined. The role of travel agents, airline schedules, tourist flows and trips, the ways in which sex tourism is constituted by both workers and consumers, and the politics associated with this, are not focused upon. Systems of provision arguably place a greater emphasis on interactions between different chains, and on commodity meanings.

Systems of provision

The global commodity chain literature has emphasized the broader political economy of links between production and consumption, whereas the systems of provision approach focuses on specific elements of the chain. Research based around systems of provision (see Fine and Leopold, 1993a) and the Francophone *filière* tradition addresses the transformations (and trajectories) of the commodities themselves.[5] It examines the systems, spatialities and social relations which shape the flow of goods and knowledges surrounding them. However, the distinction between the global commodity chain and systems of provision approaches has become increasingly blurred and there is considerable complementarity between the two approaches (Goodman, 2001), with much work on systems of provision focusing on the buyer–supplier interface (Leslie and Reimer, 1999).

Fine and Leopold (1993a; Fine, 2002) are perhaps most closely associated with systems of provision perspectives, pinpointing the ways in which production and consumption are linked through various commodities. By focusing on different sites and practices, systems of provision (SOP) incorporate the symbolic meanings, discourses and material uses of objects as they relate to transformations in commodities and the 'systems' in which they are located. While chains are conceived in a vertical and linear fashion with the emphasis on transformations at various sites along a chain, the SOP approach also accommodates interactions between horizontal factors (nodes or phases with commonalities examined across a number of commodities, for example, advertising, design, retailing) (Glennie and Thrift, 1993; Leslie and Reimer, 1999). A different sort of politics may be invoked than that indicated by GCC literature – one which is grounded in the multidimensional rather than firm oriented power relations which are constituted around the particularities of sites of production and consumption (Fine, 1993: 600).

However, like the GCC literature, systems of provision perspectives have been criticized for their continuing emphasis (following Marxist traditions) on uncovering the hidden social relationships behind a particular commodity (Goodman and Dupuis, 2002: 6). The approach is one which sees consumption determined by a complex chain of activities resulting in a minimization of the status of consumption and its transformative powers in the delineation of systems of provision (Lockie and Kitto, 2000). Thus the 'consumer emerges only to disappear again into a production centred framework' (Goodman and Dupuis, 2002: 7). Systems of provision, like work emanating from the GCC tradition, tends to be oriented towards discovering structural tendencies which are a feature of the interaction of transformations of commodities and the broader social relations which constitute states, capital flows and forms of regulation.

Systems of provision chains have provided valuable insight into commodity trajectories in spatial and social contexts. They emphasize the contexts in which commodities acquire form and meaning, as a way of demonstrating how things are embedded in social life and of exploring the social life (or commodity biographies) of things (Friedman, 1994; see Box 5.4).

Some commentators argue SOP approaches give insufficient attention to the complex practices and relationships through which production and consumption are linked (Hughes, 2000) and there is no indication, for example, of how the materiality of commodities might influence their social lives (Lockie and Kitto, 2000). Leslie and Reimer (1999) dispute this, arguing the systems of provision approach does move beyond an emphasis on economics, technological change and modes of regulation which sees consumption demand as merely shaping production, or the end phase in a chain of activities. Fine (2002: 90), in a second edition of *The World of Consumption*, reasserts the importance of the economic and material in consumption studies, arguing a systems of provision approach can adequately integrate both the material and the cultural. He suggests a cultural system is attached to each system of provision, thus addressing tensions 'between the image of the commodity and what it actually is' (2002: 90).

Thus the social relations surrounding commodities have a critical role, with consumption being translated throughout the commodity chain and not simply being located in the terminating links. Leslie's research on fashion clothing demonstrates the material and symbolic basis of production–consumption chains. The fashion chain is one in which production is highly globalized, but 'whose display, marketing and ultimate consumption is localized on the body' (2002: 62). The conditions of employment in retailing – minimum wages, quotas, strict regulatory regimes, surveillance techniques, specified 'performances', control of bodily movement and appearance – are such that female retail workers share much in common with women situated at other sites of the chain (2002: 73). Fashion clothing as a commodity influences and constructs workers' bodily subjectivities at retailing, advertising and consumption sites in the chain, blurring the line between work, production and consumption. Box 5.4 also draws on the systems of provision approach to examine 'how the distinctiveness of clothing as a commodity reverberates through the entire chain' (2002: 63).

BOX 5.4 TRANSNATIONAL BIOGRAPHIES: THE USED CLOTHING INDUSTRY

In her book *Salaula: the World of Secondhand Clothing and Zambia* (2000), Karen Tranberg Hansen examines multiple sites in the commodity chain for the multibillion dollar second-hand clothing industry, tracing clothing from its donation in the West to sorting houses in North America and Europe, and into Zambia. *Salaula* is the term in the Bemba language given to second-hand clothes and means to 'select from a pile in the manner of rummaging'. *Salaula* encapsulates the idea that what matters for the people of Zambia is not where the clothes originated from, but how people choose and deal with second-hand clothes in the context of their everyday lives.

Trade in second-hand clothing extends back as far as the seventeenth and eighteenth centuries, but since the end of the Second World War there has been substantive growth in the second-hand export market to Africa. This growth was underpinned by donations of clothes from charitable organizations, and the sale of overstock to commercial buyers. Between 1980 and 1995 worldwide exports of second-hand clothing grew sixfold in value from US$207 million in 1980 to US$1410 million in 1995 (2000: 115). Since the 1980s, Hansen argues, 'second-hand clothing is less about charity than it is about profits' (2000: 18), with an estimated 50 per cent of clothing donated to charities in the United States, for example, being onsold to the textile salvage industry (2000: 102; see Figure 5.3). Some of this material is used for fibre reprocessing and recycling of rags, so grading used clothes into what is deemed usable or not by 'end' consumers is therefore an important social/material practice in the system of provision.

FIGURE 5.3 Clothing bins in Otaki, New Zealand: part charitable donation, part commercial recycling enterprise

Much of the clothing destined for Zambia is shipped from the US and Europe to ports in Tanzania, Mozambique and South Africa. The 45–50 kg bales of clothes must clear customs where duty is paid on a per weight basis to the Zambian Revenue Authority. Bales are then sold as part or whole bales to individual traders. Lusaka is the *salaula* 'trade capital', where traders purchase clothes to onsell to other traders, exchange for other commodities or sell to 'end' consumers. The profitability of wholesalers is often determined by cultural factors including local consumer knowledge and/or reputation and the source of the stock at any given time. Traders make purchases based on knowledge about the presumed saleability and desirability of clothing, and knowledgeable practices of display and arrangement are critical in the successful trading of *salaula*. For example, clothing from the United States is viewed as more worn and of lesser quality than clothing originating from Europe.

Unlike the second-hand clothes mentioned in the case study of charity shops in Britain in Box 4.6, where clothes closest to the body are not desirable, bales of underwear, lingerie, night attire and shoes are sold. *Salaula* in Zambia is seen quite differently: it is 'new', having no history despite consumers' knowledge of it as donated clothing. To differentiate Zambian from imported *salaula* clothing, the *salaula* bales are not washed and ironed (as they might be in many First World charity shops). The public opening of bales by traders, and the folds and crinkles the clothes possess, allay suspicion that the garments were previously worn by Zambians (which is not desirable) and show that they are instead 'fresh from the source' (2000: 172).

The study of *salaula* also demonstrates connections between the cultural politics and material realities of consumers in the First and Third Worlds. Hanson suggests many First World consumers may be unaware that their donations of clothes can be part of a commodity chain where goods are redistributed on a profitable rather than charitable basis. In Zambia clothing has always been an important political issue in everyday life, with the state's import substitution policies from 1964, and from 1991 trade liberalization influencing not only employment in Zambia but also the regulation and availability of locally made and *Salaula* clothing. Local cultural politics and matters of political economy both matter to how the clothing commodities are desired, purchased and worn. *Salaula* is bought by all except the top income levels in Zambia and, compared with store bought clothes, is affordable, represents better value for money and provides greater choice. The *salaula* trade has not stifled the work of traditional tailors; instead many are involved in altering garments or recycling less desirable attire as new items such as sweatshirts, dress shorts and blankets.

Hansen touches on the different cultural and material socialities and spatialities of the second-hand clothing chain across place, but her study is also of interest because she endeavours to look beyond literature which focuses on the negative and/or exploitative aspects of global commodity chains. While recognizing the complexity of the debate, Hansen considers such effects must be interpreted more broadly, arguing *salaula* has contributed positively to Zambian life through its appropriation in Zambian material culture. Though Zambians' construction of *salaula* is informed by discourses of development which constitute 'donations by developed countries', the clothes are redefined as new by consumers, providing a medium for the production of a spatially specific cultural economy of taste and style. Hansen highlights the way in which discourses of clothing in Zambia are about circulation, appropriation and bricolage rather than emulation of Western fashions, noting how *salaula* says something about one's place in the world

> (Friedman, 1994: 112–16). Hansen's study demonstrates that the social and economic life of the commodity makes a difference to how it is consumed. She also emphasizes the importance of the commodity movements and practices to the contexts in which they are made meaningful, going beyond simply tracing the places in which chains touch down.

Both global commodity chain and systems of provision perspectives are useful for understanding how 'commodities touch upon multiple scales and pass along the chains from the body to the city and global arena' and how 'individual commodities articulate differently with these scales' (Leslie, 2002: 62). GCC and SOP research informed by Marxian political economy perspectives tends to locate the politics of power in the sphere of production rather than consumption (Goodman and Dupuis, 2002). This can have the effect of neglecting the power dynamics of consumption or of failing to adequately problematize consumption as a 'place – locational and social – in struggles for change' (Johns and Vural, 2000: 1196). Fine (2002) disagrees with this point, but there have been few studies which explore production and consumption interdependencies and how power is effected at multiple sites along commodity chain systems.

Commodity Circuits

An inadequacy of the linear and unidirectional conception of commodity chains is the implicit assumption that there is a necessary link between the various sites or nodes in the chain (Miller, 1997). By combining elements of political economy and poststructuralism in a circuit with neither beginning nor end, a circuits approach does not privilege production or consumption. A non-linear or circuits approach conceives of commodity transformations and the social practices, relationships and politics associated with them as different moments which may be contingently rather than necessarily linked in different ways. Tracing movements of commodities, people and practices through other forms of connectivity helps elide the prioritization of one 'moment of commodity circulation over another' (Hughes, 2000: 177).

Du Gay et al. (1997) use the concept of a circuit of culture (see Figure 5. 4) to understand commodity 'moments'. The circuit allows consumers to construct, interpret and make meanings though commodity linked practices at all nodes in the circuit (rather than at the consumption end of a chain, back along to the social relations of production). Transformations in commodities and translations in commodity meaning occur through processes of representation, identity, production, consumption and regulation (rather than along linear chains), with each process being linked with other processes through many interactions. Adequately studying a commodity involves following commodities as they move through and across the circuit. Though the circuit is divided into discrete sections, they overlap and 'intertwine in complex and contingent' ways (1997: 4; see Box 5.5).

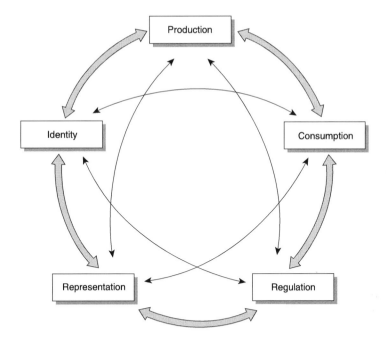

FIGURE 5.4   The commodity circuit (redrawn from du Gay et al., 1997, with permission)

BOX 5.5 THE STORY OF THE SONY WALKMAN

Du Gay et al., in their book *Doing Cultural Studies: The Story of the Sony Walkman* (1997) explain how the circuit of culture approach might be applied to the study of the Sony Walkman. This involves focusing on the multiple ways in which commodities are articulated through these processes. For example, such an approach might involve exploring the Walkman's *representation* in language and how advertising has a role in *identity* construction. These things are a part of the *production* of the Walkman as a cultural artefact, comprising how an object is produced technically and how it is accorded and makes meaning through material culture (involving the production of culture and the culture of production – circulating back through issues of 'identity and representation'). How the *identity* of the Walkman is (re-)created by Sony (the firm) is also important. This identity in turn links to consumers and the role of the media and designers in highlighting the articulation between production and *consumption*. The meaning commodities have in people's lives is an important part of consumption, but this is not unmediated, as production, consumption and the other nodes in the circuit are subject to *regulation* (1997: 4–5). Thus the circuits of culture approach advocated by du Gay et al. (1997) endeavours to rupture cleavages between production and consumption, but ironically it does so by separating and to some extent privileging culture.

> Fine (2002) criticizes the arbitrary and reductionist nature of du Gay et al.'s circuit of culture, arguing it does not sufficiently address the stresses on the cultural sphere wrought by the circuit of capital. He suggests that the cultural and material structures, mechanisms and imperatives of capitalist society are unavoidable and cannot be discounted, and that much of the research in this vein has simply 'invested sites along the circuit of *commodities* with a cultural content' (2002: 106).

Within geography, non-linear approaches have been significant. They have been utilized to examine relationships between advertising, producers and consumers (Jackson and Taylor, 1996), to focus upon sites and translations through second-hand cycles of consumption (Gregson et al., 2002b) and to study food consumption and production (Cook and Crang, 1996; Hollander, 2003). The concept of 'displacement' (Cook and Crang, 1996; see Box 4.7) draws upon the circuits of culture approach as it examines how culinary culture is constituted through commodity meanings and practices as they circulate and are reconstructed across systems or networks from one site to another.

Rather than 'circuits', many recent studies tend to use 'systems' or 'networks' as more appropriate metaphors by which to capture the discursive and material transformations and practices which surround movements of commodities and knowledges across place (though these studies are not necessarily informed by the actor network approaches outlined in the next section). In these studies 'vertical questions and dilemmas' appear to surround more horizontal socio-spatial relations of consumption (Crang, 1996: 63). The emphasis here is on the construction, expression, translation and circulation of power as it operates through circuits which link producers and consumers through various contexts. For example, Hollander's (2003) research on supermarket narratives (the stories regarding place and production that appear on product packaging) explores the politics of the filtering and regulation of food information and how the politics of food consumption is shaped by numerous contradictions with regard to how sugar narratives (such as ethical, organic or green discourses) are produced, accepted and interpreted by consumers.

Campbell and Liepins (2001) used a non-linear approach to researching 'organic' food. They explored it as a discursive field, comprising the competing meanings 'by which the world is understood, reconstructed – even governed' (2001: 25). This involved an understanding of how commodities interact through sites, and the multiple ways in which meanings were transformed. This enabled them to explore not only how the notion of 'organics' is socially constructed (for example, through the formation and interpretation of organic standards, global food scares and social movements) but also how this reshapes production and consumption contexts (such as farm, commercial arena) and the processes and forms through which meanings of 'organics' are constructed (through growers, inspectors and corporate influence).

Another non-linear perspective is evident in studies of transnationalism (see Box 6.5) which examine 'the traffic in things' (Jackson, 1999), emphasizing transformations and connections between sites and moments of consumption and production. By researching the routes commodities take as they flow within commodified and non-commodified processes, and the construction of their meanings through various cultural intermediaries, Jackson (1999), Dwyer and Jackson (2003) and Crang et al. (2003) trace the processes by which the identities of firms, commodities, personnel and consumers are negotiated and constructed. Their work on 'ethnic' foods and fashions has highlighted the ways in which commodity circuits are leaky and ambiguous.

Actor Network Theory

In actor network theory (ANT) the metaphor of a 'network' is used to depict more complex interrelationships than those represented by the metaphor of a chain (or even circuits of culture), enabling commodity linkages to be realized as 'webs of interdependence' and relationships (Hughes, 2000: 188). Like commodity chain and circuits literature, the actor network perspective is not a unified theory but a series of approaches which seek to explore how human and non-human actors are connected.

Drawing on poststructuralism and developed by Bruno Latour, John Law, Michel Serres and Michel Callon (Law and Hassard, 1999; Serres and Latour, 1995), one of the key features of ANT approaches is that the distinction between the economic and cultural spheres is blurred, because the social world is viewed not as a series of territories but as a circulation held together by things (Latour and Woolgar, 1986). An important aspect of this is a revised concept of 'agency'. Rather than agency being vested in sentient human beings with intentionality, it is understood as the 'collective capacity of material associations and relational networks' to act or have effects (Lockie and Kitto, 2000: 11).[6] People, entities, nature, objects and discourses are viewed symmetrically, with each potentially possessing transformative capacity within the networks they are situated (Murdoch, 1997a). The separation between nature and culture is refuted as both are seen as outcomes of network building processes rather than pre-existing ontological categories (Goodman, 2001).

Latour (1999) is careful to state that the network should not be conceived of as structure, with the people as agency. Even human agents are always networks because all human activities are types of circulation (including corporeality, subjectivity, thinking and moralities) (Law, 1994). Actors or actants may be non-human, e.g. animals, machines or matter. A music CD might be said to be an (enrolled) actant which has a role in evoking physical, emotional and spiritual effects in the listener. Thus actantality is not what an actor does (such as listening to music) but what provides actants with their actions, subjectivity, intentionality and morality (Latour, 1999). With ANT, consumption must be viewed as being constituted by the relational settings in which it occurs, with actantality determined by relations established within the network (Murdoch, 1998). The act of listening to music, for example, is not independent of the functionality of the stereo, its location, the material and social production of the CD, the body–mind through which the sound circulates and is felt, other people and 'other things' in the room (or elsewhere), or even previous experiences of listening to music. The actor network is the act, therefore, linked together with all its influencing factors.

An ANT conceptualization bypasses some of the dualisms that characterize considerations of scale in human geography and in commodity chains (non-ANT approaches to understanding scale are discussed in Chapter 3). While interactions may be localized, they are always to some extent constituted by distant actions (Murdoch, 1997b: 321). Scale is conceptualized not in terms of levels of representation (such as global, national, local, body), but rather in terms of 'how things are "stitched together" across divisions and distinctions' (1997b: 322). The key issue, therefore, 'is not one of scale but of connectivity' (Suryanata, 2002: 72).

Networks are always local, and so can be understood only in specific time–place contexts, and as globalized only in terms of their physical extension across space in practice. Latour (1993) illustrates this with the example of a railway which is local at all points, but also global because it stretches across countries. Networks pleat and fold 'space–time through the mobilizations, cumulations and recombinations that link subjects, objects, domains and locales' (Murdoch, 1998: 357). 'Spatial analysis is also network analysis' so the task of the researcher is to follow networks along their lengths 'localizing and globalizing along the way' (1997b: 332, 334).

Scale represents the outcome of the heterogeneous links established between actors (1998: 321). Heterogeneous associations can be conceived as a multiplicity of connections which are fashioned out of a diverse range of materials, relationships and activities which are constantly 'becoming' (note the similarities to how identity formation is described in Chapter 7) rather than already constituted (Thrift, 2000a: 5). Entities, people and things come to be enrolled, combined and disciplined within these networks in ways which are performative, with actorial roles emerging as effects rather than causes (Murdoch, 1997b). Networks endure when material resources and conditions and actants build stable social relations in which the circulation of 'immutable mobiles' occurs, e.g. devices, types of people, animals, money, which can be transported from one location to another without changing form (Thrift, 2000a: 5). Mediaries and intermediaries do the work of keeping networks connected and folded into each other (2000a: 5). While immutable mobiles help networks to endure, networks are not fixed entities but a shifting series of transformations and deformations.[7] Discourses operate as modes of ordering which also establish relations between actors, entities and places in the form of coherences, patterns and performative relationships (Law, 1994).

Questions of power can be directed towards understanding how some local centres in the network permit, constrain or enable actors to influence the shape/position/form of others at a distance, a type of remote control (Murdoch, 1998). Thus a network analysis would see producers in developing countries and retailers and consumers in the developed world as the product of 'complex flows between a whole host of interconnected actors that have become enrolled in the network' (Hughes, 2000: 178). Hughes' (2000; 2001) research on the global cut flower trade provides an excellent illustration of how this might work. She demonstrates how the power of retailers to influence trade actually hinges on their capacities to adapt complex networks in their favour and to circulate particular kinds of knowledge (managerial and techno-scientific knowledges, and knowledges which pertain to purchasing, flower care, aesthetics of display, cultural meanings of flowers and gift giving).

Network approaches have been used as an organizing framework to 'map commodity-chain co-ordination and, through this, to comment upon the social constructions of spaces, places, and territories' (Pritchard, 2000: 790; see, for example, Smith, 2003 on world city actor networks). Hitchings (2003) takes a different approach, and in doing so highlights the performance of power which occurs in a single connection in a network (see Box 5.6).

BOX 5.6 PLANTS AND PEOPLE: AN ANT APPROACH

Hitchings (2003) looks at private gardening, focusing on one aspect of an actor network – plant and human relationships. Starting with plants as actants, Hitchings notes the plants perform themselves into existence as discrete entities, with the gardener's role being to coax out plant performances (2003: 107). When he began his network exploration with people, plants became passive entities; gardens were understood as feats of human creativity and people performed as garden designers. Hitchings' study may seem banal, but it provides an important insight into how objects are lived with, and how power can oscillate through networks, shifting between plants (gardener as plant person) and people (gardener as designer) (see Figure 5.5).

FIGURE 5.5 Hitchings' ANT study demonstrated how power can alter through a single association in a network, oscillating between plants and people

ANT has the potential to provide a different set of insights into the study of consumption because of its assumptions about the constitution of networks. The first is the avoidance of *a priori* assumptions of agency by actors/actants or nodes (compared with commodity chain, systems of provision, circuits of culture literature). The second is the emphasis on connectivity and mobilities as causal factors in the becoming of networks (a feature of circuits approaches, but without the prescription of the nodes which comprises a network). The third (and perhaps most controversial) is the actantality which is accorded symmetrically to nature, animals and inanimate objects. Much of the geographical work utilizing ANT approaches has been with studies of agro-commodity networks. ANT research on food networks provides some insights into how a politics of food might be combined with the politics of production (Goodman and Dupuis, 2002: 15). Box 5.7 examines Whatmore and Thorne's (1997) ANT research on coffee networks, and Smith's (1996) discursive analysis of Starbucks cafés. Each of these studies provides a partial perspective on the complex relationships between consumers and producers, articulating the possibilities of a politics of consumption in different ways.

BOX 5.7 LINKING CONSUMPTION AND PRODUCTION: BLACK BOXES AND COFFEE CONSUMPTION

Smith (1996) suggests the history of coffee is bound up with the rise of capitalism and European colonialism, establishing relations of domination and dependency both between Europe and its (ex) colonies and within them. The contemporary coffee system of provision reifies relationships of dependency in connecting developing nations of the South to the rich industrialized nations of the North. Processing and distribution are dominated by large trading companies and multinational food corporations. The segment of market controlled by the top five coffee producing companies in Europe, for example, is as high as 77 per cent (Renard, 1999: 494).

Whatmore and Thorne's (1997) ANT study explored the development of a UK consortium called Cafédirect which sources, imports and markets the Cafédirect brand of fair-trade coffee (given institutional legitimacy by a hybridized form of itself). They examined Cafédirect partners in Peru, an exporting cooperative named CECOOAC-Nor, which comprises nine individual coffee producing cooperatives. Whatmore and Thorne found that discourses of partnership, alliance, responsibility and fairness were performed in very different ways to those which characterize commercial coffee networks. Though alternative coffee producers used some of the same spaces, practices and disciplines of the market as non-fair-trade coffee producers (for example, contract deadlines, stock exchange, just-in-time rationales, processors), the mode of ordering of enterprise in fair-trade networks was mediated by modes of connectivity which differed substantially from the 'cost-minimizing, self interested individual of neo-classical economic theory' which connects those who grow and buy the coffee (Whatmore and Thorne, 1997: 298; see also Tan, 2000).

The fair-trade coffee network is constituted through the performance of fairness which applies to a fair price for growers and an excellent product for consumers. Coffee is an important actant in the network: low quality coffee beans, for example, mean the commodity will not be suitable for fair-trade networks and will result in its sale to commercial networks. The encouragement and enhancement of organic farming practices is also a feature of the mode of ordering of connectivity which could be used in the marketing of the fairly traded commodity.

Whatmore and Thorne's (1997) study makes explicit the significance of the associations (relationships, objects) connecting producers and consumers, but the potential role of consumers, of coffee consumption, and of coffee consumption sites was not examined. Despite the possibilities of heterogeneous associations as a means of understanding the connections between people, things, practices and relations, ANT approaches have tended to emphasize the sphere of production and institutional actants within it. It is this 'black boxing' of consumers which some have argued has been a continued feature of both commodity chain and ANT approaches (Lockie and Kitto, 2000).

In contrast, Smith's (1996) examination of the cultural practices associated with coffee production examined how local consumption practices, spatialities, discourses and politics are linked to the political economy of coffee production. By investigating discourses of coffee drinking he demonstrates how Starbucks, through both its marketing and its store information strategies, has transformed coffee drinking from a 'taken for granted act of consumption into something resembling a hobby' (1996: 509). Starbucks refashions, appropriates and redistributes elements from historic and contemporary economic geographies of coffee production in order to invest symbolic (and saleable) meaning in the commodity and in the consumption of it (1996: 515). For example, the company has invented the Starbucks Passport, to enable its American consumers to drink their way around Starbucks' imagined and essentialist 'Third World' of coffee regions. Smith concluded that Starbucks marketing strategies deliberately emphasize (particular and partial) relations of production, making them part of the commodity itself; they remove from consumers' site/sight the structures of domination and exploitation that underpin commodity production (such as labour conditions).

Though Smith (1996) has focused on the consumption end of the coffee commodity chain and highlighted the potential powers vested in discourses of marketers and retailers through the partial explication of relations of production, his study also tells us nothing of the transformative power of acts of consumption. Smith willingly acknowledges his study has only described the subject positions that were made available in discourses of Starbucks, not the ways in which consumers respond to them. Whatmore and Thorne's (1997) study also touched upon consumers as actants in networks but without a sense of the complexity or constitution of this group.

How best, then, to overcome the problem of the 'black box' of consumption? Smith argues for the 'undertaking of ethnographic studies in one location and connecting

> them with other moments which are part of the same process'. It would seem that this might provide a way forward, a means not just of connecting consumption and production processes, practices, modes of ordering and discourses in space and time, but, of following the flows, circulations, translations and deformations of knowledges, discourses, practices, and human and non-human entities through commodity networks. Nevertheless, discussing the significance of consumption in the commodity chain and network approaches remains a dilemma for geographers.

Concepts of governance, relational power and the operation of discourse have also been adapted in ANT literature. Networks become integrated configurations of governance mechanisms (Lewis et al., 2002), with discourses operating as modes of ordering to help establish their durability (Law, 1994). Power is a relational effect and operates through 'remote control' via lines of flow and heterogeneous connections in the network (Whatmore and Thorne, 1997: 290). It is manifest in successful attempts to enrol other actants into a network through a process of 'translation' in which the interests of others are aligned and mobilized as part of that network (Lockie and Kitto, 2000: 8). Thus it is the capacity to act rather than to wield power which is significant (Goodman and Dupuis, 2002: 18). However, the capacity to invoke change through a network appears much more remote and diffuse than in the chain or circuit approaches, with the potential abandonment of a language and standpoint around which people can mobilize (Leslie and Reimer, 1999).

Nevertheless, Goodman (1999) suggests ANT has the potential to form a radical politics of nature by refuting the modernist ontology of a separation between nature and culture. In emphasizing how things are held together, network approaches allow for the existence and visibility of new forms, practices and spaces of resistance through a non-hierarchical and non-essentialist conception of actants (be they human or non-human). ANT can provide a more nuanced and complex understanding of production and consumption relations through the 'intricate interweaving of *situated* people, artifacts, codes and living things and the maintenance of particular tapestries of connection across the world' (Whatmore and Thorne, 1997: 289).

The development of poststructural political economies (Larner and Le Heron, 2002a; 2002b) which draws on actor network theory and the concept of spatial imaginaries (Larner, 1998) represents a further attempt to understand how production and consumption relations are 'shaped up'; are given form, legitimacy and power (Lewis et al., 2002). Emphasizing a poststructural concept of governance, this literature focuses on 'how we think about governing others and ourselves in a wide variety of contexts' (Dean, 1999: 18) rather than on the 'who or what' of government. Geographers advocating this perspective argue shifts in discursive formations influence the restructuring of political and economic discourses and spaces (for example, through calculative practices of auditing and benchmarking). Though primarily focused on exploring globalization, poststructuralist political economies have the potential to provide insights into specific consumption processes and discourses through examining 'how relations amongst people and things might be imagined, assembled and translated to effect co-ordination at a distance' (Larner and Le Heron, 2002a: 417; see Box 5.8).

BOX 5.8 SHAPING MOTHERHOOD AND MILK CONSUMPTION: POSTSTRUCTURAL POLITICAL ECONOMIES

The imaginings of 'motherhood' constructed in Sri Lanka by the New Zealand Dairy Board (NZDB) and the Movement of Mothers to Combat Malnutrition (MMCM) provide some insights into how this approach might be applied with regard to consumption. Alison Greenaway, Wendy Larner and Richard Le Heron (2002) demonstrate how differing representations of mothers as consumers of milk by the NZDB and the MMCM were central to the emergence of the market for milk powder in Sri Lanka and to how motherhood was shaped. The NZDB saw itself as meeting the needs of mothers in providing (scientifically valued) nutrition, trustworthy products and food security; while the MMCM discouraged consumption of purchased milk powder, supporting the empowerment of women and promoting self-sufficiency as a means of confronting malnutrition. Both institutions focused on mothers as key actors in addressing the problem of malnutrition, but the NZDB saw mothers as active autonomous consumers, making informed choices about milk consumption, while the MMCM portrayed mothers as passive victims of globalization, vulnerable to commodification, and needing to breastfeed in order to sustain motherhood and achieve active social transformation. Tensions of hegemonic discourses linking nutrition and consumption, and resistance and empowerment, shape the market for milk in Sri Lanka, but do so in a context in which discourses that do not shape the commercialization of milk are hidden. Ironically, neither the MMCM nor the NZDB, in representing mothers as consumers of milk, acknowledge women's role in the dairy industry as informal and small scale milk producers (Greenaway et al., 2002).

Greenaway et al.'s (2002) research demonstrates that power geometries emerge from the patterns of centrality and marginality which comprise networks. Tracing lines and flows of associations which comprise networks can therefore become a key to understanding how power is 'manufactured' (Murdoch, 1997b: 335). The conception of actants, rather than human agents that drive and configure the networks, is what differentiates Massey's concept of power geometries (Box 1.11) from those of ANT theorists.

Power, Politics and Connectivities

In this chapter I have outlined a number of perspectives which conceptualize consumption not as a temporal and spatial act, but as a sphere of social relations which is created and reproduced across a variety of scales and times. Instead of a momentary act, these perspectives demonstrate how consumption is a placed, material, imagined, connected and shifting series of cultural and economic processes. I have focused on three different modes of connection; chains, circuits and actor networks. Each of these implies different standpoints of power, and hence different assumptions about the ways in which the politics of consumption and production might be constituted. Yet each of the perspectives has provided geographers with differing insights into how commodities move and are made meaningful across space.

Global commodity chains have highlighted the unevenness of flows of commodities along production and consumption chains, particularly focusing on their role in relation to processes of economic globalization. Fine and Leopold's (1993a; Fine, 2002) systems of provision approach demonstrates how relations of connection differ depending on the commodities and the structures and histories in which they are embedded. The work on commodity circuits highlights nodes through which commodities circulate and take on meaning. This has the advantage of focusing on alterations across horizontal facets of commodity relations, and has enabled consideration of how commodities and knowledges about them may travel and be translated and transformed through particular sites. Actor network approaches are able to examine the complexity of commodity relations, and may emphasize how processes at a distance play a role in governing relations between actants (be they human or non-human), objects and the places through which relationships are constructed.

Chain, circuit and actor network approaches have begun to touch on the sorts of spaces created through modes of connection, but few studies have been able to focus on multiple sites through which production and consumption relations are constituted to explore the subjectivities and socialities involved, whether up and down chains, via circuits or following commodity networks. A central concern remains the mapping out of the spatialities of power, including how best to trace shifting lines and geometries of power (Leslie and Reimer, 1999) and how to establish a politics of connection (Hartwick, 2000). More insight is needed into the consequences of particular power geometries for transforming existing social relations between actors, actants, things and places. It also appears that work on the (modernist) regulation of consumption, particularly by states and local governments, and research in the area of collective consumption, remain removed from much contemporary theorizing of consumption (though see Clarke and Bradford, 1998; Fine, 2002; Pacione, 2001; Preteceille, 1986). The regulation, governance and framing of consumption as practice and discourse (for example, through specific actants such as state, firms, local government, NGOs, commodity chain actors, consumers) needs to be explored in depth. This would enable further understanding of the connections between socialities, spatialities and subjectivities of consumption. Institutional regulation remains a strong feature of much research on retailing but appears to be less predominant in relation to other aspects of consumption. It is important, too, to broaden understandings of connections between production and consumption beyond food or horticultural commodities which have been the focus of much work in the approaches discussed in this chapter. Nevertheless, the approaches outlined have made significant inroads into exploring the complexity and vibrancy of connections between place, things, people and commodities.

FURTHER READING

Crang, P. and Cook, I. (1996) 'The world on a plate: culinary knowledge, displacement and geographical knowledges', *Journal of Material Culture*, 1: 131–53.

du Gay, P., Hall, S., Janes, L., Mackay, H. and Negus, K. (1997) *Doing Cultural Studies: the Story of the Sony Walkman*. London: Sage.

Gereffi, G. (2001) 'Beyond the producer-driven/buyer-driven dichotomy', *IDS Bulletin*, 32 (3): 30.

Hartwick, E.R. (2000) 'Towards a geographical politics of consumption', *Environment and Planning A*, 32 (7): 1177–92.

Hughes, A. (2000) 'Retailers, knowledges and changing commodity networks: the case of the cut flower trade', *Geoforum*, 31: 175–90.

Hughes, A. and Reimer, S. (2004) 'Introduction' in A. Hughes and S. Reimer, S. (eds), *Geographies of Commodity Chains*. London: Routledge. pp. 1–16.

Leslie, D. and Reimer, S. (1999) 'Spatializing commodity chains', *Progress in Human Geography*, 23 (3): 401–20.

Murdoch, J. (1998) 'The spaces of actor-network theory', *Geoforum*, 29 (4): 357–74.

Whatmore, S. and Thorne, L. (1997) 'Nourishing networks: alternative geographies of food', in D. Goodman and M.J. Watts (eds), *Globalising Food: Agrarian Questions and Global Restructuring*. London: Routledge. pp. 287–304.

NOTES
1 Raikes et al. (2000) argue the notion of a commodity chain emerged from Immanuel Wallerstein's (1974) world systems theory. Analysis of contemporary society from a world systems perspective would involve a consideration of the world

as a system, as a capitalist world economy in which changes in and development of space in a particular place are linked to political-economic decisions connected with the accumulation of capital, creating a continually changing core, periphery and semi-periphery. Commodity chains form the basis of a commodity system in which particular divisions of labour are manifest; these are regulated by the cyclical expansion and contraction of the capitalist world system over time.

2   It has been suggested that these campaigns are connected with broader anti-globalization movements concerned with such issues as multinational control, international trade and finance, the environment, Third World debt and poverty (Mercier, 2003).

3   Silvey (2002), in discussing university protests against Nike workplace conditions and her own role on a University of Colorado committee charged with investigating the implications of a code of conduct for producers of logo-bearing apparel, notes how the discipline of labour was not simply a feature of the supplier end of the chain. The corporatization of the university also meant that the disciplining effects of transnational corporations extended to the consumption end of the chain (through threatening to withdraw contract funds).

4   There is some debate over whether fractured politics qualifies as political action (see Goodman and Dupuis, 2002: 5, for differing opinions on food politics).

5   The Francophone *filière* tradition has been applied primarily to agricultural commodities and is derived from a variety of research traditions and thoughts. The concept of *filière* revolves around building a physical flow chart of commodities and their transformations and, like the systems of provision research, includes the creation, structuring and operation of power that surround specific commodities (Raikes et al., 2000). Goodman (2001) argues labour process is a central conceptual foundation of *filière*.

6   While some have claimed ANT addresses the structure–agency debate, Latour (1999) argues his intention was never to occupy a position in that debate, or to overcome this contradiction, but was to bypass it by paying attention to these things and following them elsewhere through processes of circulation.

7   Phil Crang's conception of displacement is also relevant in thinking about ordering and connectivity of mobilities, addressing Mike Crang's (2002: 571) call to explore geographies of mobility, interstitial spaces of social life, spaces of movement, transit and passage, and spaces of brevity and ephemerality (and the need to see such spaces as having their own effects and significance).

6

Commercial Cultures

In the previous chapter three ways of exploring connections between consumption and production were discussed. Geographers have endeavoured to look beyond the broad contours of connections to the imbrication of culture, production, economics and consumption and their mutual constitution in space. Within geography, research on commodity and commercial cultures has been at the forefront of challenges to consumption and production as dichotomous categories. This chapter examines commercial cultures by exploring three case studies: music, the serial repetition of McDonald's, and indigenous tourism. These case studies reveal how assemblages of people, entities, are brought together in place, and how commercial cultures shape and position subjects in different ways. The concept of commercial cultures also enables one to critique views of globalization as producing homogeneous and unidirectional consumption experiences and spaces, and to understand how flows of knowledge, commodities and people are caught up in transnational spaces.

Understanding Commercial Cultures

Culture comprises the social practices of everyday life through which people create meaningful relationships, experiences and places. Mort suggests 'cultures of consumption are the point where the market meets popular experience and lifestyles on the ground' (1988: 215). The term 'consumer culture' is usually a reference to groupings of consumers based on propensity to consume, consumption practices, lifestyle or objects of consumption. Rather than inferring consumption is only a 'cultural' matter, we may use the term 'commercial culture' (see Box 6.1) to encapsulate the linkages between consumption and production, and the inseparability of production and consumption, commerce and culture, the material and the symbolic.

> **BOX 6.1 COMMERCIAL CULTURES**
>
> Over the last two decades consumption geographers have endeavoured to find ways of reconciling opposing ways of seeing and overcoming dualisms between the economic and cultural, symbolic and material (see, for example, Clarke and Purvis, 1994; Gregson et al., 2001b). In a book entitled *Commercial Cultures* editors Peter Jackson, Michelle Lowe, Daniel Miller and Frank Mort advocate research which explores how aspects of cultural production 'are inherently concerned with the commodification of various kinds of cultural difference' (2000: 1). Research on how hybrid 'commercial cultures' emerge also forms a way of understanding how the market and processes of commerce are embedded in a variety of cultural processes. Tracing the connections between economies, practices and spaces provides a means of exploring commercial cultures (2000). This involves understanding the particularity of practices, flows, networks and relationships and their movement, assemblage, displacement and permeability through temporal and spatial contexts.

The interdependence of culture and economy has been addressed in numerous studies, for example, McDowell and Court's (1994) research on service cultures, and Bryman's (1999) discussion of how 'Disney culture' pervades service industry delivery. However, studies which recognize the inflection of the cultural in the economic, or the economic in the cultural, may restate rather than overcome the culture–economy divide. Gregson et al. (2001b) suggest that culture is often used as an additive in the explanation of phenomena and activities which are seen as primarily economic, as in studies of cultural industries or in the culturalization thesis – the idea that 'economies are increasingly about the production, distribution and consumption of items that are cultural in character' (2001b: 621). If culture is about the production of meaning and the processes and practices which comprise routinized ways of doing and seeing, then for 'economy and cultural to be conjoined, meaning and practice must be regarded as inseparable' (2001b: 630). However, the challenge remains as to how to reconcile such binaries without privileging either side.

Examination of commercial cultures must be underpinned by a sense of their spatial constitution but also via an exploration of how power operates and is distributed (see Chapter 5). Jackson (2002b: 15, citing McRobbie, 1997: 85) argues a politics of consumption must involve thinking across material and symbolic realms to identify anxieties and tensions which provide opportunities for political debate and social change. The discussion on music which follows examines how 'cultural' questions of aesthetics, taste and style cannot be divorced from 'political' questions about power, inequality and oppression (Jackson, 1993: 208).

Music: Exploring the Politics of Commercial Culture

Commercial music provides fascinating insights into how meanings are constructed and power is articulated through cultures, economies, practices and spaces. The purchase, use and experience of music delineates boundaries between self and others, providing a means of thinking about spatiality and the politics of its production. Music is also a site of social struggle: discursive and material practices surrounding the production and consumption of sound are gendered, racialized and classed (Revill, 1998) and music is a sphere in which the politics of First and Third Worlds and the intersection between global and local is constituted (Kong, 1995: 190).[1]

Music is transgressive, with sounds and meanings of music being 'linked to geographical sites, bound up in our everyday perceptions of place, and a part of the movements of people, products and cultures across space' (Connell and Gibson, 2003: 1). Music has been used in strategies of political socialization and the constitution of patriotism or national identity (Smith, 1994) and as a component of 'place-making' through the promotion of cultural industry, tourism and local economic development strategies. Music plays a significant role in the construction of particular narratives of the local. For example, the illegal use of industrial warehouses for house dances in Blackburn, England (and associated consumption of the drug Ecstasy) established new and transgressive soundscapes in the late 1980s and early 1990s, invoking tensions between partygoers/organizers and owners of the premises, police and local residents, politicians and local authorities (Ingham et al., 1999).

While much geographic attention has been focused on analysis of the lyrical contents of songs and their textual representation (Kong, 1995), 'sound itself has an important and active role to play in the organization of social, economic, and political spaces' (Revill, 2000: 597). Accordingly music should be considered a means of communication in itself rather than just a medium for some other communicative practice, for example, as a vehicle for lyrics or as a channel for social interaction and display (2000: 597). An emphasis on music as performance rather than text shifts the methodological and theoretical focus towards how geographies are embodied, experienced and made rather than simply 'read' (S.J. Smith, 2000).

If music is constituted powerfully through its production, then this is also a characteristic of its consumption. The ways that listeners themselves 'perform' has an influence on the political, economic and emotional spaces of music as 'listening makes music too' (2000: 634). Music is experienced as a performance of power enacted by consumers which is creative and productive, bringing spaces, peoples and places into form. The music of the Australian band 'The Whitlams', for example, both reflected and helped constitute a radical politics opposed to inner city change in their home town, Newtown, Sydney (Carroll and Connell, 2000). The subsequent formation of a 'community' of Whitlam fans provided a focus for subordinate groups to negotiate the dominant power system. Thus music can play a role in processes of identity formation and modes of personal expression, particularly with regard to youth subcultures (Chatterton and Hollands, 2002; Wallace and Kovacheva, 1996).[2]

Music comprises a pervasive, sometimes insidious component of the geographies of everyday life (Kong, 1995). While music is consumed directly through choices to attend concerts and nightclubs, and to purchase and listen to CDs, cassette tapes and videos, it is also consumed as part of the production of other consumption experiences and sites. Music played in a restaurant, for example, may provide ambience and 'mood' or be interpreted as 'noise' when it detracts from the consuming experience. Music is part of knowing and being in place and is consumed actively and passively in public and private spaces (Frith, 1996; see Box 6.2).

BOX 6.2 INSTORE MUSIC: THE PRODUCT AND PERFORMANCE OF CONSUMPTION

The privileging of the visual in studies of shopping malls and retail spaces (see Chapter 3) has meant a relative neglect of the pervasiveness of sound/music and its potential significance in consumption (DeNora and Belcher, 2000). Tia DeNora's and Sophie Belcher's (2000) research drew on Goffman's concept of performance (see Box 4.4) to observe and interview consumers, employees and store managers in order to examine the role of music in encouraging feelings commensurate with the displayed commodities. They discovered music played in stores ranged from music directed to the presumed identities of shoppers, to that which served to 'brand' items for sale or to heighten the exclusivity and specificity of sales product and place (2000: 93). Music provided part of the 'scene' in which producer–consumer performances were enacted. Shoppers were observed toe tapping, moving to music, and singing. Store music also encouraged an emotional investment on the part of staff in relation to the commodities on offer, becoming a tool of surveillance for management, with head offices of retail companies determining the timing and content of music to be played and in some cases ensuring store operators could only play company music in stores. Music in retail outlets not only facilitates a tactile and bodily experience of consuming, but forms a means of establishing the social contexts of commodities and the regulation and surveillance of employees. Retail music is consequently both performative and about performance.

DeNora and Belcher's research indicates how commercial cultures which characterize contemporary music are intricately connected to cultural politics and the production of place. However, commercial cultures are constituted through a web of relationships which do not begin or terminate with the musicians or consumers (S.J. Smith, 2000). Changes in software technologies and Internet distribution systems, for example, are transforming networks of creativity (involving creation and interpretation of music through multiple acts of performance), reproduction (dealing with manufacture of multiple copies of works), distribution and consumption (involving both retail organizations and consumers) (Leyshon, 2001: 60).

Digital technologies which allow the copying, dissemination and consumption of audio files over the Internet, and the downloading of files (primarily as compressed MP3 formats) to personal computers, have also begun to alter the material form of music networks and the commercial cultures which constitute them. MP3 and similar formats have been welcomed by some music industry agents who believe they will reorient power relations away from retailers and back to musicians and artists. Others view such techniques as a potential form of piracy which will pose a threat to sales and existing forms of ownership (and exploitation) in the music industry (2001: 52).

Tensions are emerging as a consequence of changes in the relationship between those institutions and agents involved in the production of music commodities and those involved in facilitating their consumption (such as between recording companies and major retailers). Dislocated relationships may exist between those whose expertise is based on technologies and techniques of cultural and textual production, and those whose expertise is based upon knowledge of the consumption of cultural artefacts and services (du Gay and Negus, 1994).

In addition, despite discourses of freedom, agency and expression which circulate around consumption of music, consumer choice is often scrutinized and regulated by retail management. Du Gay and Negus (1994: 411) argued that routes of consumption 'lead directly into (and out of) the design offices, marketing departments, boardrooms and assembly plants of electronic, communication and media corporations'. Executive decisions of recording companies, for example, are based around assumptions of marketability of music arising from existing and presumed consumer preferences. Information technologies, and products like SoundScan (a point-of-purchase technology) which tabulate sales to consumers, contribute to power/knowledge being vested in companies which distribute music, enabling control over the range and nature of products available to the consumer and the level of profits (McCourt and Rothenbuhler, 1997). Point-of-sale technologies also facilitate backward integration, in which retailers may have greater control of product design, development and other processes (du Gay and Negus, 1994: 397). The structure of the retail music sector, in which a small number of retail companies operate multiple outlets, has had implications for industry relationships and the commodities available and accessible to consumers. Independent stores are less economically productive for large record companies to deal with, so they are often forced to specialize in niche music (du Gay and Negus, 1994).

Thus debates over the production/consumption of music, how the industry is structured and the formats in which consumers have access to audio 'bytes' involve a politics and a geography of consumption in which relationships between producers and consumers are both blurred and articulated through complex networks. The case study in Box 6.3 explores alliances and tensions which operate as part of the commercial culture which comprises hip-hop music.

BOX 6.3 HIP-HOP: RAP AS RESISTANCE?

The incorporation of hip-hop music in commercial cultures which operate through formal spaces of market exchange is not divorced from the cultural politics of consumption. Hip-hop and its most recognized component, rap, are said to have formed 'on the street' and in the urban neighbourhoods of 'black America' in the 1970s. The sounds and lyrics of hip-hop represent both a description of everyday life and desires, and a politicization centred upon disaffection, resistance and rage in the context of a diaspora of the 'black Atlantic' (Gilroy, 1993). Hip-hop's politics are also a politics of style as the music embodies creative and aesthetic volatility as part of its stylistic heritage (Neal, 1997). Rap involves the consumption of and creative appropriation and reconstitution of existing sounds, images and technologies, missing, cutting and scratching these as 'artifacts of locally based cultures', often through the selective reinterpretation of the music products of big business (Connell and Gibson, 2003: 183).

The commodification of rap has meant it has also become concerned with the creation of identifiable and marketable products and 'intellectual' properties (Negus, 1999). Though major label companies have bought independent rap companies and allowed them to function relatively autonomously in order to maintain their assumed 'authenticity' (Connell and Gibson, 2003), Negus (1999) notes how the corporate organization of the music industry has also meant hip-hop has been marginalized. This marginalization retains the mythology of hip-hop as radical black music from the street, but creates a commercial culture which is 'central to the changing business practices and aesthetics of the contemporary music industry' (1999: 492). In the USA, the segmentation of black music into notoriously unstable divisions of major record companies has often occurred with less economic investment than other types of music, an absence of senior African-American executives above the 'black' division in the music hierarchy, and assumptions of a short 'shelf life' for the music. The assumed costs of incorporating others' intellectual property in producing tracks, and payment structures in the industry, have all meant that hip-hop and rap music has not had a (powerful) corporate space in which to establish its own agenda (Negus, 1999). Though hip-hop and rap can be constituted as a lifeline of hope for diasporic peoples, a key question is how such music can function as a viable place of working as well as thinking, one in which economic livelihoods are possible and not just desired (McRobbie, 1999).

Despite the potential for corporate 'containment' and shaping, new and radical forms of hip-hop have continued to emerge. A continual redefinition of rap as the music has crossed social and cultural barriers within and outside the USA has occurred despite, rather than because of, the ways in which the recording industry has organized the production of popular music (Negus, 1999: 504). French and German rap, for example, bears similarities to rap which has evolved in the USA but also has important differences with regard to national and racial issues, exemplifying different challenges to the structures of the societies in which they are located (Connell and Gibson, 2003).

The commercial viability of hip-hop and the potential to diffuse its message of social critique were largely contingent on the success of the new digital technologies which it tacitly critiques (Neal, 1997). Emphasis on hybridity (see next section) and the forging of new ethnic identities in such music can too easily ignore how categories are easily packaged and commodified by the record industry marketing these sounds 'as a new take on the exotic otherness of marginal cultures' (McRobbie, 1999: 148). While the globalization of hip-hop can invoke radical resistance and opposition to forms of racism and marginalization, its politicization is tempered by commercialization (Miles, 1998b).

Commercial cultures are always in a sense hybrid, mixed and remixed differentially in place (Qureshi and Moores, 1999). Bennett's (1999) research on the consumption of hip-hop music by white Newcastle (England) residents illustrates the differing discourses and politics which emerge in place. Where hip-hop was seen as 'authentic' black music, its enjoyment and consumption were articulated around a politics which invoked a strategy of distinction by its consumers, separating knowledgeable consumers whose political affinities recognized black politics and/or who saw affinities with their own working class experiences. These consumers constructed others' appropriations of black music as inappropriate, unknowing and inauthentic (in much the same ways as 1970s retro wearers constructed those who engaged in carnivalesque modes of dress: see Box 4.6). Other consumers and producers of hip-hop in Newcastle saw it as a unique expression of being in place, as part of the actual construction of what it meant to be working class in Newcastle by its consumers (Bennett, 1999). Similarly hip-hop in Zimbabwe, Italy, Greenland and Aotearoa/New Zealand is not simply an appropriation but is constituted through complex modes of indigenization and syncretism to create a vehicle for the expression of indigenous resistance vernaculars, their local politics and the 'moral' geographies associated with them (Mitchell, 2000: 52).

Understanding the commodification of cultural practices such as music, and how actors and discourses in situated contexts frame commercial cultures, thus provides a means of understanding how consumption operates as a spatial and social phenomenon and the geometries of power which are associated with it. The discussion on music as a commercial culture has also illustrated that 'cultural' questions of aesthetics, taste and style cannot be divorced from 'political questions about power, inequality and oppression' (Jackson, 1993: 208). A key question in thinking about consumption of music is the extent to which consumers can construct their own meanings in the context of music produced for them by the music industry and how constant negotiation and flows between commerce and creativity in music create a complex geometry of power (Miles, 1998a).

FIGURE 6.1 The global homogenization paradigm is linked to Americanization of place (Herb cartoon, permission by Chris Beard)

Consuming and producing music are very much about politics, embodying asymmetries of power which result from the ways in which music commodities are made, marketed and sold, the naturalization and prioritization of some views/values above others, and how 'cultural' meanings of different groups such as recording artists, companies, retailers and consumers (such as what constitutes rap and rap consumption) are assembled, transformed, contested and reaffirmed across space. The next section explores these ideas in relation to processes of globalization, considering how best to understand flows of people, things and commodities which comprise commercial cultures.

Commercial Cultures and Globalization

Globalization is itself a contested concept, but generally seems to involve a sense of the increasing interconnectedness of the world, and the extension and deepening of social relations and institutions across space (Amin, 2002). With regard to consumption, globalization appears to have been linked primarily with concepts of cultural homogenization, cultural imperialism and/or Americanization (see Figure 6.1). Research on commercial cultures critiques this idea, arguing that globalization is simultaneously cultural and economic, and that places are not simply initiators or recipients of global objects, processes and knowledges.

Global homogenization

Howes (1996) argues contemporary thinking about the cultural effects of the circulation of commodities has been dominated by the global homogenization paradigm. This perspective views globalization as the convergence of global culture. Global homogenization results, with the recolonization of spaces via the market, the replacement of local goods by mass produced commodities usually originating from the West, and the subsequent erosion of cultural difference and diversity. Transference of material objects and the practices, meanings, images and symbolism associated with them has also been associated with deterritorialization – the severing of 'original' identities, signs and meanings from traditional locations (Short et al., 2001). The global homogenization thesis also cites as evidence the marketing and distribution practices of companies with global branded products like Coca-Cola, Nike and Levis, despite these companies only producing a small proportion of branded goods (Miller, 1995).

Ritzer's (1993) theory of McDonaldization is often seen as symptomatic of global convergence. The concept of McDonaldization is based around the notion that the principles of the fast food restaurant pervade the organization of production and consumption on an increasingly global scale. Ritzer (2002) argues that the preoccupation with rationalization through securing efficiency, calculability (through quantifiable properties of object and labour), predictability and control now extends to a wide range of other contexts, including education, family, legal and justice systems and Christian churches. However, Ritzer did not cleary differentiate between McDonaldization as a set of rationalizing principles and the geographical extension of McDonalds (Smart, 1999). Ritzer's theory provides provocative reflections on the nature of control in contemporary organizations, but it is too simplistic to view the worldwide growth in McDonalds outlets in terms of global homogenization. With the opening of new restaurants different commercial cultures have emerged (Figure 6.2) and as Box 6.4 demonstrates, the geographic extension of McDonalds embodies aspects of both cultural and economic homogenization and heterogenization.

FIGURE 6.2   The changing nature of public space. The first McDonald's in Russia opened in Moscow in 1990, this is one of approximately 103 restaurants currently operating in the country

BOX 6.4 MCDONALD'S: CONSUMING MEANINGS IN SACRED AND THERAPEUTIC LANDSCAPES

Maoz Azaryahu discusses the conflict that arose after the opening of a McDonald's restaurant adjacent to a national military shrine at Golani Junction in Israel in 1994. The opening of the restaurant disrupted the meaning of Golani Junction as a memorial space. As a site of remembrance for those who had lost and continue to lose their lives as part of Golani Brigade (a prominent and active military unit in Israel), Golani Junction was a shrine of bereavement and patriotism, a sacred space of community (1999: 482). The opening of McDonald's at Golani Heights produced a 'monument' but of quite a different kind, acting as a metaphor of 'an alleged cultural transformation of Israel society and, in particular, for the Americanization of Israel' (1999: 486). Opposition to the arrival of McDonald's was based around the assumed culture of the restaurant chain rather than its utilitarian function as a place of food consumption (which was lacking on the site). McDonald's function as a site of social rather than culinary pleasure was juxtaposed with the sombre sacredness of the adjoining memorial. Pressure exerted by bereaved parents and a direct intervention by Yitzhak Rabin (Prime Minister and Defence Minister at the time) resulted in a compromise with McDonald's officials. McDonald's undertook to close the restaurant when memorial ceremonies were held, and the visual (and symbolic) prominence of the Golani memorial site was restored by architectural modifications and planting aimed at concealing the restaurant.

However, as a symbol of Americanization the arrival of McDonald's in Israel was interpreted in complex ways. It was seen by many as detrimental to local Israeli traditions, particularly by religious Jews who saw it as symptomatic of a secular society and its negation of dietary laws (Azaryahu, 1999). McDonald's (more than other American owned fast food chains such as Burger King and Subway) represented the American way of life, one which did not necessarily evoke negative sentiments for many Israelis. McDonald's was also viewed as signifying Israel's integration into a 'homogenized, global consumer culture' (1999: 485).

Robin Kearns and Ross Barnett argue the arrival of McDonald's in New Zealand (in 1976) was also a 'potent symbol of the multinational colonization of the New Zealand landscape' (2000: 89) but since then has become a ubiquitous part of many New Zealand children's everyday lives. It was the opening of a McDonald's restaurant within the Starship hospital, New Zealand's major tertiary teaching children's hospital in Auckland, which was controversial, occurring within 'a broader narrative of the commercialization of health care in New Zealand' (2000: 81).

In 'boldly going where no health enterprise had gone before' (2000: 84) as in Israel, it was the meanings and moralities associated with McDonald's which were a source of controversy. Operating McDonald's in the Starship was seen by some as an endorsement by health professionals of the acceptability of takeaway food, the moralities of which were particularly questionable in a region where dietary problems and poverty related illness are prominent. The second was based on the perception that the deal

> forged in 1997 between McDonald's and management represented the 'predatory actions of the private sector on an ailing public health system' (2000: 86). Responses in favour of the development were based around pragmatism, quality and convenience, and getting the best service for the hospital. Given that potential benefactors of the Starship are consumers of a branded image, McDonald's was part of the extension of symbolism of 'hospital as mall' for consumers, but also an intrinsic part of the saleable Starship metaphor (2000: 90) with McDonald's assisting in the imageability and marketable nature of the hospital as a product for potential sponsors. As in Israel, McDonald's was required to make various concessions: minimum external signage, nutritional extras on the menu, and two nationwide health awareness campaigns a year. Kearns and Barnett (2000: 90) argue that somewhat ironically the franchise is extending the rationality and efficiency of processes of McDonaldization in hospital care while simultaneously providing consumers with a fantasy laden eating experience in an enchanted space.
>
> The cases of the McDonald's outlets in Golani Junction and the Starship hospital demonstrate the widely variant discourses which surrounded the opening of the new franchises. In Golani Junction, McDonald's conflicted with moral geographies surrounding remembrance, patriotism and sacred space, yet for others its arrival reinforced political and cultural alliances. In the Starship, debates were located within discourses of consumerism and commodification of healthcare, healthy eating and the formation of a friendly and normalizing therapeutic landscape. Experiences of eating thus became entangled in the debates about the commercial culture of the commodity being provided. Understanding such debates involves considering the economic and cultural (re)production of McDonald's, and the ways in which the consumption and production of meanings, imaginings, commodities and experiences are transformed in place and across space.

Thus even for a firm such as McDonald's, which has become an 'icon of global homogenization of landscapes and culinary tastes' (Azaryahu, 1999: 481), the cultural homogenization thesis may be critiqued. Places and people are not passive and receptive agents of an (unchanging) culture or cultural object which originates from elsewhere (Miller, 1998). McDonald's has altered the production, processing and marketing of commodities in particular localities, and people have responded to and shaped the McDonald's experience differently in place (a point reflected in Watson's 1997 volume *Golden Arches East* exploring McDonald's in Asian countries).

Creolization, hybridity and transnationalism
A second paradigm to do with understanding how consumption objects and cultures change across borders is connected with 'creolization'. This occurs

when material objects cross borders and 'the culture they "substantiate" is no longer the culture in which they circulate' (Howes, 1996: 2). Creolization focuses on the inflow of commodities to a place (as opposed to the outflow from the First World, America or the West as suggested by the homogenization paradigm) to consider how commodities are assigned meanings and uses. Creolization emphasizes how locals selectively appropriate elements of the receiving culture in order to construct their own hybrid medium (Barber and Waterman, 1995). Culture becomes creolized as a consequence of the fusion of disparate elements which are both heterogeneous and local. The creolization paradigm stresses the active, creative and experiential role of people as cultural producers rather than passive recipients and is not premised on a notion of an authentic and potentially corruptible culture.

However, though the creolization paradigm sees cultural change as dynamic and fluid, it still relies on the notion of a 'receiving culture' where commodity meanings appear to be unchanged in processes of transference. This implies the origins of change are always 'outside' a bounded space, and processes of change are homogeneous and static in their (external) constitution. Thus the concept produces a binary which constructs culture into indigenous (traditional and local) and imported aspects (modern, global) (Barber and Waterman, 1995). While creolization allows for consumer creativity and the significance of local processes, like the globalization paradigm it retains an emphasis on the passivity of the consumer (Jackson, 1999).

Bhabha's (1994) notion of hybridity which has been linked with creolization is of greater use when focusing on the interface between objects and the commercial cultures which surround them. Hybrid cultures are 'much more than syncretisms; they are not mixtures from two or more sources, but a creation of something new out of difference' (Shurmer-Smith and Hannam, 1994: 139). Hybridity involves places of encounter and is constituted through the boundaries between cultures and cultural flows; it is the passage, 'the boundary, being the place where something begins its presencing' (Bhabba, 2001: 140). As a site of 'newness' the middle place/passage may be a site of displacement and disjunction, syncretism, juxtaposition, redefinition and (re)creation (2001: 140). Hybridity in this sense refers to things and processes which transgress and displace divisions between same and other, for example displacing and re-creating meanings of 'the West and the rest' which tend to underpin notions of cross-cultural consumption.[3]

Homogenization, creolization and hybridization are partial ways of seeing the world and not necessarily mutually exclusive. Nederveen Pieterse (1995) argues that the other side of cultural hybridity is transcultural convergence. Thus while forms of hybridity may emerge, there may also be differential experiences of homogeneous processes and new kinds of particularism. Similarly, the concept of 'reterritorialization' rather than deterritorialization might better encapsulate the ways in which cultures, identities and socialities take place and form through processes of transformation away from traditional locations and origins (Short et al., 2001). Transformations in commercial cultures may as a consequence be destabilizing or enabling, they may be contiguous across space or polarizing. Recent research on transnationalism is beginning to address the nature of hybrid formations of commercial (or commodity) cultures by tracing what happens to people and products as they move across space (see Box 6.5).

BOX 6.5 TRANSNATIONALISM AND COMMERCIAL CULTURES

Research by consumption geographers on transnationalism avoids assumptions of bounded authentic places or cultures of origin to examine the mutual constitution of local and global by tracing the people and products of transnational commercial culture. This involves exploring how commodity differences and meanings move (Jackson's 1999 'the traffic in things'), and how they are produced, consumed, translated and displaced across space in situated contexts (Crang et al., 2003).

Using a non-linear approach (see Chapter 5) this research on transnationalism explores how people in embedded spaces engage with other objects, capitals, people and knowledges and how these constitute transnational space as a 'multidimensional space that is multiply inhabited and characterized by complex networks, circuits and flows' (2003: 441). Research on British based women's wear firm EAST and Indian based company Anokhi demonstrates how the notions of cultural difference are produced and consumed in contrasting ways transnationally. This includes consideration of how both firms imagine 'India' and of how difference is designed materially and symbolically. Anokhi, for example, emphasizes local craftsmanship and distinctive production techniques such as block printing (somewhat ironically commodifying ethical and sustainable enterprise in India), while EAST creates an ethnic or exotic look that is not place specific. Dwyer and Jackson (2003) also examined the way in which consumers in Britain negotiate and transform meanings and value, demonstrating similar tensions to the ways in which producers imagine India. They conclude that the commercial culture is an ambivalent space, noting that both EAST and Anokhi are involved in the dynamic and fluid production of ethnic difference. Consumers are viewed as actively constructing meanings of self, others and transnational commodities in spaces which are themselves 'caught up' in transnational networks (Dwyer and Jackson, 2003). The case study of Anokhi and EAST demonstrates the approach can adequately encapsulate both the materials and economies of the movement of goods and services and an appreciation of the commodification of cultural difference (Jackson, 1999: 105), demonstrating too how differently positioned producers and consumers inhabit transnational spaces (Jackson, 2002b). Studies of consumption and transnationality thus seek to develop an explicitly geographical understanding of commodities and the cultures which surround them.

Understanding the powerful effects of production and consumption networks in a globalizing world is not simply about content or the 'what' of change (as in the arrival of McDonald's or the influx of new commodities, knowledges and practices) but is also about the form of change, that is the means by which people communicate differences to each other in ways that are 'more widely intelligible' (Wilk, 1995: 124). Studies of transnationalism are beginning to unpack this, examining how relatively stable processes (such as fabric design: see Box 6.5) are actively constituted, powerfully promoting some differences while subsuming others.

Tourism: Consuming Culture as Other?

Exploring tourism as a commercial culture can enable one to consider how people communicate differences to each other through practices of production, representation, identity formation and consumption. A number of studies of tourism and 'cross-cultural' consumption have focused on the representation of indigenous cultures as the 'other' and the appropriation of indigenous cultural practices, images and artefacts (Scheyvens, 2002). Yet commercial cultures may produce hybridized spaces and presences which are not exploited appropriations of the other vested in notions of authentic and static 'cultures'.

Tourism, like other social processes, can only be understood in the historical, spatial and social context in which it occurs. Tourism as a consumption process may involve aspects of the mundane but also fantasy and escape, capitalizing on real and imagined differences between familiar and unfamiliar places. As a commercial culture, tourism is a conduit for organizing meaning in space (Hughes, 1998) and is bound in complex cultural politics which links producers and consumers of tourist 'products' through such media as guidebooks, advertisements and tourist brochures and participation in tourist practices such as photography, sightseeing, tourist trails and package tours.

Discourses of colonialism, imperialism, racism and gender can underpin how commercial cultures of indigenous tourism are represented and manifest. A thrust of tourism promotional literature often portrays indigenous peoples as members of primitive, static, unchanging societies, removed from relationships which characterize the 'modern' world (Morgan and Pritchard, 1998). Said (1978: 1) notes that the Orient, for example, provides one of the 'deepest most recurring images of the "other"' in which the East is portrayed as a place of escape' (from the West) and fantasy. The Pacific, like the Orient, has been similarly exoticized, with New Zealand being viewed as a kind of 'Pacific Arcadia' (Hall, 1998). Marginalization may occur as a consequence of the powerful tourist gaze leading to alterity, the construction and state of 'otherness' which arises out of the difference between self and others.[4]

New Zealand/Aotearoa presents a particularly interesting case study (Box 6.6) within which to study tourism as commercial culture. While tourism is about construction of difference, in its commodified form it is also about the creation of economic value. Economically tourism is significant, being the second largest export earner behind the dairy industry, contributing $14.6 billion to the New Zealand economy, or 9 per cent of GDP. In addition, debates surrounding the representation and practice of tourism as a commercial culture cannot be separated from processes of identity construction and the hegemonic power relations which exist between Pakeha (non-Maori people primarily of European descent) and Maori (indigenous dwellers of Aotearoa).

Since British annexation in 1840 and a second wave of European colonizers (the first occurring in the eighteenth century) the traditions, customs and histories of Maori were 'reinterpreted into a new cultural and historical framework by the knowledge "keeps" of these new migrants' (McGregor and McMath, 1993: 45). Nineteenth century colonization brought disease, conflict, land confiscation and a subsequent loss of economic livelihood. While there has been a resurgence in population in recent years, with one in seven New Zealanders (526,281 out of the total New Zealand population of 3.8 million) identifying themselves as Maori, individuals with Maori ancestry are under-represented in statistics on employment and income but over-represented in unemployment figures (27.7 per cent of all unemployed people in 1996: Statistics New Zealand, 2001). Unfortunately social statistics tend to present Maori as a form of underclass, obscuring positive aspects of Maori identity, masking *iwi* (tribe) and *whanau* (family) differences and the active participation and contribution of Maori in everyday life in Aotearoa. Yet material differences and inequities do matter; political-economic factors in tourism, for example, provide impediments to access for training and education for tourism related careers and make securing finance more difficult (TheStaffordGroup, 2001).

Since the 1970s a cultural renaissance in Maoridom has occurred, centred on representation of Maori, and issues of autonomy, land and self-determination.[5] Sissons (1993) indicates that the cultural renaissance must also be understood in relation to state practices which have brought about the systemization of tradition involving the enhancement of the state's image as a representative of both Maori and Pakeha interests. Systemization of Maori tradition has resulted in 'ethnicization', the selective appropriation of beliefs, values, practices and places which has resulted in the fracturing and objectification of Maori culture, with the result that Maori culture has become a symbol of ethnic and indigenous distinctiveness and a strategic state resource (Sissons, 1993). This has important implications for the construction and representation of Maori within national tourism strategies, and for the commercial cultures which establish the 'indigenous tourist product'. Understanding the commercial culture of tourism in NZ as a hybrid transnational formation enables one to move beyond representing 'Maori' as an object to be consumed to understanding how consumption is produced through commercial cultures with particular outcomes for people and place (Box 6.6).

BOX 6.6 ONE HUNDRED PER CENT PURE? MAORI AS THE WELCOME PARTY OF AOTEAROA/NEW ZEALAND

'Maori are sick of being the welcome party of New Zealand Tourism' (Maori Affairs Minister Parekura Horomia, quoted by Espiner, 2001: 2). Mr Horomia's statement was a call for Maori to be owners and investors in tourism operations rather than simply cultural 'performers'. That commerce and culture are inseparable is exemplified in the practices and discourses of Maori tourism in New Zealand/Aotearoa. Under-representation of Maori employees, low investment levels and a lack of financial resources are issues for Maori tourism as much as issues of representation involving the cultural packaging and marketing of 'things' Maori (TheStaffordGroup, 2001).

FIGURE 6.3 Maori as the Welcome Party of New Zealand: a tourist advertisement sponsored by the New Zealand Government Tourist Bureau and Tasman Empire Airways Limited [1953] (permission by NZ Tourist Board and image supplied by Alexander Turnbull Library, Wellington, New Zealand)

It was in the 1870s that tourism first became formalized in Aotearoa/New Zealand. Since that time, until the early 1980s Maori were represented in tourist brochures, postcards and guidebooks as 'noble savages' living in an exotic land, whose spiritual existence was romanticized and lifestyle eroticized, adding 'background colour and uniqueness to the national tourism product' (McGregor and McMath, 1993: 45; see Figure 6.3). Yet Maori were not without agency during this process. The identities of some *iwi* (tribes) in the latter part of the nineteenth century were strongly connected with their role in tourism (Ryan and Crotts, 1997). A number of Maori women guides of Te Arawa *iwi* during the colonial period also became 'a considerable economic and political force' (Taylor, 1998: 2).

It is only in the latter half of the twentieth century that Maori financial involvement in tourism developed (Barnett, 1997). There is a continued consumer demand for 'Maori tourist product', with 93 per cent of tour operators selling inbound tour packages that contain some form of Maori tourism such as *hangi* (earth cooked meal), a Maori 'concert party', a tour of Maori thermal reserves, villages and so forth – practices which are seen to typify an authentic 'Maori culture' (TheStaffordGroup, 2001: 19). The popularity and demand for activities indicate nothing of consumer experience of indigenous tourism, but for travellers from both Asian and European countries a desire for authenticity becomes important, encountering rather than just gazing upon 'indigenous people' in settings which are 'real' and not too contrived (Tourism New Zealand, 1995).

In recent years the New Zealand Tourist Board's marketing has been founded on its 100 per cent Pure campaign which focuses on representations of pristine, clean, green and spectacular environment but also hyperbole around adventure tourism and real cultural experiences of place. The campaign was criticized for cultural insensitivity over the use of images involving Maori to promote the country overseas and for its under-representation of Maori (Espiner, 1999). Efforts have since been made to incorporate a more contemporary focus to Maori culture by including a virtual *powhiri* (welcoming ceremony) on the website and by downplaying the use of the facial *moko* (tattoo) which the organization admits led some overseas tourists to assume they would see Maori with facial tattoos everywhere in New Zealand (Espiner, 2001).[6]

Issues of 'authenticity' in tourism have also been debated by Maori. Some suggest a tacky and insensitive 'plastic *poi*' culture has emerged – a referent to the use of plastic *pois* (small balls which hang from string swung in women's dance) and plastic *pi-pius* (skirts traditionally made from flax) (Sell, 1999). Others argue it is not the plastic *pois* which are the issue but the embodied performance of *kapa haka* (performing art) and the context in which this occurs (Murray, 2000). Tamaki Tours, a highly successful Maori operated tourist company, for example, has created an alternative to the romanticized archaic Maori image by creating 'a living village experience' which seeks to provide visitors with experiences that are 'authentic', emotional and spiritual, that present traditional and contemporary cultural perspectives developed in consultation with tribal elders to respect traditional protocols (TheStaffordGroup, 2001: 114).

However, the contested nature of 'appropriation' and 'commodification' is not just framed within New Zealand. In 2002 the British Broadcasting Corporation shot a new opening sequence for its BBC1 programming which featured a collection of dance scenes, including a *haka* performed by a Welsh rugby team.[7] Maori lawyer Maui Solomon believes it is ironic that the BBC 'as an icon of colonialism' chose the *haka* to rebrand itself (Quirke, 2002b: 1). Jo Hutley, the Maori Londoner who taught the *haka* for the BBC1 introduction, argued that he had the right to teach 'his' culture to anyone, and that the use of the *haka* in this way was appropriate because it was undertaken in the context of education of participants about the cultural and spiritual significance of the *haka* performance (Quirke, 2002a: 3). The debate about the legitimacy of 'cultural practices' performed out of context is fascinating because it concerns judgements not only about how practices will travel and be translated as 'commercial' cultures, but about the very 'authenticity' of their productions in situated place. *Haka* specialist Pita Sharples, for example, finds pop singer Robbie Williams' arm tattoo (which derives from a particular *iwi*) more offensive than 'fun' based *haka* adaptations because he believes this is a direct appropriation of tribal intellectual property (the Maori tattooist disagreed) (*The Dominion*, 2001).

In recent years notions of hybridity and transnationalism have been implicitly acknowledged as part of discourses around commercial culture. Maori scholar and artist Darcy Nicholas, for example, believes Maori as part of the modern world are freely using symbols, materials and concepts that belong to 'other' world cultures (*The Dominion*, 2002: 19). He draws on concepts of hybridity rather than 'appropriation', suggesting because the *mana* (prestige/authority/power) of *moko* and *haka* has become diluted through their uses as decoration and entertainment, this has made these cultural practices/forms 'part of the world of design and art and all the creative possibilities beyond that' (2002: 19). New Zealand fashion designer Charles Walker also believes Maori art is living, vibrant and changing. He suggests the production of artefacts and art forms is based on a syncretism where there is no boundary between traditional and modern, where past symbols and signs are used in untraditional fields (such as contemporary fashion, media or pottery) and where traditional fields (such as carving) use modern techniques and search for ways to make their art new (Steed, 2000: 22).

Thus debates over intellectual property rights and tourist practices and representations represent more than power struggles over representations of Maori artefacts and practices; they signal concerns over material culture and how it is transformed across space through commercial cultures. 'Authenticity' not only becomes about the context in which these commodities, peoples or their representations are expressed, but is complicit in questions of how identities and cultures are constituted in place which is open to transnational flows of people, commodities, information and ideas. Most critically it reflects debates about essentialist versus more fluid understandings of what culture is, and how it is meaningfully constituted in place.

The case study in Box 6.6 has explored the 'varied, complex and often chaotic processes which characterize tourism' (Squire, 1998: 93) and the myriad of social practices which comprise indigenous tourism as a commercial culture. Economic and institutional barriers to ownership and control of Maori tourism operations;

systemization by the State which 'locates' and essentializes Maori culture; representations of indigenous tourism which position Maoridom as a static spectacle or which reduce culture to a 'cultural performance': all are underpinned by constellations of power which position 'actors' in multiple ways, making notions of 'cultural appropriation' inadequate as a basis for interpreting change.

The appropriation of Maori culture is a complex issue, complicated by the conceptualization of Maori culture in terms of essentialized understandings (often bound in equally indefinable notions of tradition and authenticity), its presumed separation from other commercial cultures, and the need to articulate ownership in commodity terms such as 'intellectual property' or 'tourist products'. If culture is about everyday struggles in specific social contexts rather than signs or symbols, then Webster (1993: 12) believes only the appearances of a culture can be appropriated, with 'the real thing' remaining in the hands of those whose culture it is. But the question remains as to what 'the real thing is', and whether it can ever sit outside those aspects of culture which are appropriated by and subsumed in other commercial cultures (the '*Ka Mate*' haka as a prelude to a rugby test, for example, has become a symbol of national identity) (Murray, 2000).

Without implying the essentialist starting points that hybridity implies, the concept of transnationalism (Crang et al., 2003) does provide a way to understand the differing meanings and subject positions that emerge through tourism as 'places of encounter'. Transnationalism offers possibilities for conceptualizing the simultaneous existence of forms of power/knowledge associated with Maori commercial cultures as means of expression, communication, identification and livelihood (for example, the simultaneous existence of institutionalized protocols for recognition of 'authentic' Maori art forms and the commensality and benefits which also derive from mutual sharing of cultural images across social and spatial boundaries within and outside New Zealand).

The case study in Box 6.6 also highlights the need to understand the role of consumers and producers in performing transnational commodity cultures. The limited amount of research on consumer preferences and experiences only begins to explore how embodied tourist experiences are represented, created and consumed. Forms, practices, people and relations of production (concert parties, or visits to living villages) and consumption (participation, photography, video recording, buying souvenirs, sending postcards, purchasing artefacts) combine to produce differing discourses of 'Maori' and 'Pakeha' which operate across spaces, and to change them even as they are performed in situated contexts. The constitution of authenticity varies depending on where such 'commodity' performances are located, whether within New Zealand or outside. Indigenous tourism as a discourse tends to separate Maori culture from the culture of everyday life in New Zealand/ Aotearoa, a separation which works to perform binaries in ways which are productive of cultural politics which may be both enabling and disabling. Thus a paradox of tourism for Maori is that while discourses of indigenous or 'cultural' tourism might constitute gazing at the other (a desire for the exotic, different, authentic), they provide a means to 'achieve legitimacy in the struggle for political and economic recognition' (Ryan and Crotts, 1997: 900).

Commercial Cultures as Connections

Following Jackson et al. (2000), this chapter has explored connections between economies, practices and spaces in order to understand how commercial cultures are articulated through music, through expansion of fast food outlets and through the production and consumption of indigenous tourism. The complex networks of production and consumption which comprise commercial cultures are grounded in particular temporal and spatial contexts through which commodities, people and knowledges flow. Recent work by geographers on transnationalism has the potential to provide critical insights not only into the formation, translation and circulation of commercial cultures but also into the fluid and relational nature of space as it is constituted through particular assemblages of practices, discourses and relationships between people, things and place. Its emphasis on the spatial constitution of people and object centred relationships provides a substantive critique of more static representations of globalization through the creolization or homogenization approaches. Exploring the complex operation of commercial cultures enables one to make sense of the fusions, transformations and power geometries that occur, forming places of presencing, alterity and encounter.

FURTHER READINGS

Amin, A. (2002) 'Spatialities of globalisation', *Environment and Planning A*, 34 (3): 385–99.

Azaryahu, M. (1999) 'McDonald's or Golani Junction? A case of a contested place in Israel', *Professional Geographer*, 51 (4): 481–92.

Connell, J. and Gibson, C. (2003) *Sound Tracks: Popular Music, Identity and Place*. London: Routledge.

Crang, P., Dwyer, C. and Jackson, P. (2003) 'Transnationalism and the spaces of commodity culture', *Progress in Human Geography*, 27 (4): 438–56.

Jackson, P., Lowe, M., Miller, D. and Mort, F. (eds) (2000) *Commercial Cultures: Economies, Practices, Spaces*. Oxford: Berg.

Kearns, R.A. and Barnett, J.R. (2000) '"Happy Meals" in the Starship Enterprise: interpreting a moral geography of health care consumption', *Health and Place*, 6: 81–93.

Smith, S.J. (1994) 'Soundscape', *Area*, 26 (3): 232–40.

NOTES

1 For example, racialized discourses of music by black composers such as Samuel Coleridge-Taylor construct this music as closer to nature and therefore more primitive and bodily than music composed by the rational bourgeoisie (and white) Western minds (Revill, 1998).

2 The concept of a 'subculture' (see Hebdige, 1979), which exists in relation to a parent culture as a medium for the appropriation and transformation of material objects, has itself been critiqued. The notion of a subculture suggests that concepts of appropriation and transformation are framed (always) in relation to a parent culture from which they are derived (Bennett, 1999). Bennett argues that a view of subcultures suggests a coherence around and an implied collective basis to consumption practices (1999: 4). As an alternative he argues Maffesoli's (1996) concept of consumer tribes and Shields' (1992a) 'neo-tribes' are more valuable. The concept of 'tribes' recognizes the fluidity and instability of cultural affiliations which are a characteristic of late modern societies, thereby providing a basis for understanding how sites of social centrality and sociality become the focus for an individual's consumption practices (rather than belonging to or stemming from a collective grouping *per se*) (Bennett, 1999).

3 Doreen Massey (1999) suggests that hybrid spaces are not neutral spaces but are associated with particular power geometries which may be empowering and/or disempowering (see Chapter 5 on regulation).

4 One could argue that my discussion of Maori, as a Pakeha New Zealander (of European descent), also involves a process of 'othering', marking Maori as definable objects of research, as those to be spoken for or about rather than with. This is certainly not my intention.

5 This has involved such things as the formation of *kohanga reo* and *kura kaupapa* (preschools and schools in which Maori culture and language are fostered), increased emphasis on *taha* Maori in education generally, greater media coverage and legislative changes to redress land alienation and related grievances caused by Pakeha, measures to reduce socio-economic disparity between Maori and Pakeha, and increased visibility of Maori protests and land occupations (Webster, 1993).

6 The *moko* is a tattoo in which a special chisel was used to excise lines on the skin of the face (the forehead to base of the chin for men, and the chin and lips of women). 'The original intention of *moko* was to inherit the symbols that belonged to one's ancestral lines and the particular skills carried within a person' (Darcy Nicholas, cited in The *Dominion*, 2002: 19). Relatively few Maori today bear a tattooed *moko* but the symbol has been used internationally for promoting products, for example by fashion designer Jean-Paul Gaultier on some of his models (Clarke, 1999).

7 The *haka* is often performed at national events, sports events and tourist shows but also at many other Maori and non-Maori community functions such as welcoming ceremonies on the *marae* (the meeting area of a subtribe or *hapu*), prize giving ceremonies, funerals and weddings (Murray, 2000: 350). The debate appears to be framed in terms of the context in which the *haka* occurs, the understanding (or lack of understanding) of the practices, meaning and significance of the performance, and the intentions of those participating in the *haka*.

7

Moralities

Throughout this text we have encountered numerous approaches to understanding consumption. These ways of seeing, knowing and representing the study of consumption in geography have been constructed through my own situated and partial perspective. Writing an undergraduate text on consumption is performative: it re-cites and brings into being particular representations of consumption, implicitly privileging published academic knowledge and the geographies of those knowledges. Yet as a published book, an object in circulation, this book may possess actantality, circulating and being translated in ways which one can neither predict nor control.

The various means of interpreting and approaching consumption outlined in this book are also performative: these ways of seeing, knowing and doing geographies have effects by 'making differences', by enacting realities which bring to the fore certain social relations, practices and discourses and subsuming others. Such research performances are central to 'how a debate gets cast and enacted, to what gets said by whom and how, and to what does not get articulated' (Gregson et al., 2001b: 617). Consequently it is important to understand the contexts and senses in which consumption knowledges are produced/consumed and to consider the moralities and politics which are associated with them.

Performativity: Seeing, Doing and Becoming Geographies

Hitchings argues that the way we think about the world is always to some degree 'informed by the capacities and properties of particular things that surround us in this world' (2003: 102). His statement shows how knowledge is linked to being and doing things in place. The argument that knowledge is situated and performative is not new[1] but is worthy of exploration. Conceptualizing consumption knowledges as performative provides a means of considering 'What work can they do?' and 'What work do they do?'[2] – a way of thinking about how different perspectives highlight particular socialities, spatialities and subjectivities.[3] The objects and subjects of consumption study are constituted in contexts which are relational, with both researchers and readers being complicit in their constitution (see Box 7.1).

> **BOX 7.1 CONSUMPTION KNOWLEDGES: PERFORMING THE SUBJECT OF CONSUMPTION**
>
> Actor network theories draw on concepts of relational power, in which actants and things are not preconstituted and there is nothing which predetermines how the kinds of things circulate and change. However, the network metaphor is itself performative, as it is re-cited through application in social-spatial contexts. For example, consider what comes to mind when you think of the word 'network'. For many readers the term might immediately conjure up images of the Internet or the World Wide Web. Thus it may be hard to conceive of the notion of a network independently of other things which we know or think of as a network. Bruno Latour, one of the early advocates of actor network theory, states he no longer likes the concept of a network exactly for this reason. 'Network' in his view has become performative of discourses of transportation without deformation – akin to popular conceptualizations of the Internet and something quite different from the fluid, circulating entity based on translation and transformation which he first envisaged (Latour, 1999).
>
> This simple illustration demonstrates not only the theory dependence of observation (Olsson, 1980) but also how the relational (and situated contexts) in which production–consumption relations are interpreted matter. The forms by which consumption–production relations are assumed to be constituted (e.g. as chains, systems of provisions, networks, commercial cultures) will as a consequence imply different possibilities and outcomes for how consumption is manifest, and how it is interpreted in place.

Thus perspectives on consumption can become what the perspective supposedly discovers, for example social-spatial relations 'become' a chain, an assemblage of commodity relations, or a bundle of heterogeneous human/non-human relations. Different perspectives on consumption attend to different sets of social-spatial relations, and consequently different geographies (Sayer, 2003). Thinking about how consumption is performed as a geographical subject provides a means of recognizing the powers, limitations and possibilities implicit in different ways of seeing and doing geography. This includes a consideration of where and how geographical discourses and practices of consumption are located, embodied, embedded, performed and travelled but also efforts to understand the types of powers and knowledges (formal/informal, academic/personal etc.) that are derived from differing perspectives on consumption, who/what is included and excluded and the (moral) judgements made about this.

Thinking critically about these questions could provide a way of overcoming what Purcell (2003: 318) identifies as 'islands of practice' where researching and writing form a 'tradition' that limits researchers' ability to analyse, 'heed or hear' phenomena outside the focus of research, even prohibiting critical analysis of the focus itself. Conceptualizing consumption perspectives as performative entities offers insight into how things and knowledges themselves are continually becoming in their capacities to do different sorts of work and to effect different power geometries, and how in turn these 'actions' reverberate across time and place (Sack, 1992). This provides the possibility of reconciling different and seemingly 'opposing' perspectives by considering the work they can do (their capacities and dispositions to assemble, translate and flow) and the work they indeed do in their (embodied) performance through research, writing and teaching. Global commodity chain perspectives, for example, are disposed to assembling and making visible institutional connections, to connecting production and consumption in linear and vertical relationships, and to exploring power as it is effected across chains, thereby enabling possibilities of transformation via consumer resistance. In contrast, work on the cultural politics of food, for example, has made visible forms of powers neglected in political-economic studies, considering how social relations form a domain in which meanings are constructed, thus enabling one to look not just at how power might be wielded but at how it might be shaped up, negotiated and contested. The two perspectives both provide valuable insights into consumption but differ in their objects of study, their capacities to examine power relations, and the ways in which political transformation might occur.

This text has endeavoured to address some questions about the work different perspectives do, and their capacities to assemble people, things and processes in different ways to understand how consumption operates through space. Moving beyond a discussion of the merit and limitations of particular epistemologies to thinking about how such knowledges are performed, travelled, transformed or re-cited in contexts involves not just understanding how ideas are situated, but considering how they are transformed as they move from, to and across spaces. Thus it becomes important to consider questions like 'For what reasons, from what places, and to what result, have political-economic perspectives produced consumption as an effect?'; 'What does the framing of forms of consumption as "alternative" or "mainstream" do?'; 'What possibilities and limitations are implied when isolated parts of consumption processes are studied, for example, exchange, shopping, use, or disposal?; 'How might other starting positions (for example, beginning with the commodity, the consumer, relationships between entities, experiences or events) challenge, add insight to, limit and enable understanding?'; and 'Out of what (and whose) contexts, experiences, discourses have particular consumption perspectives arisen?' These questions become issues not only of knowledge production, but also of the circulation and consumption of knowledge (Desforges and Jones, 2001).

Identifying absences in geographies of consumption is important because, as well as signalling silences, they point to the uneven way in which knowledges are produced and consumed across space. The hegemony of Anglo-American geography and the English language (Garcia-Romon, 2003) means that other

FIGURE 7.1   Consumption outside Anglo-American geographies? A food market in Singapore. Critical praxis in consumption geography must be open to the experiences, voices and imagings of others as they construct discourses, materialities and practices of consumption 'in their own terms'

stories and ways of seeing the world may not be as visible, particularly for those (like myself) whose language and positionality are oriented toward these spaces of power.[4] The uneven production and consumption of knowledge have real consequences, for example, emphasizing research undertaken in spaces of the First World and examining the experiences, practices and places of working and middle class consumers.

Consumption is marked by significant inequalities, within nation-states and globally. Consumption practices for many in the First World are not focused on lifestyle or identity choices, and can operate as a source of social exclusion (Williams et al., 2001; see Box 7.2). However, consumption outside the First World is often framed as lack and seen less as a productive and meaningful sphere of social life (though see Boxes 2.7 and 5.4: Miller, 1988; Miller and Slater, 2000). Literature on consumption practices in developing countries appears to constitute a separate 'island of practice' often published outside geography journals (for example, Osella and Osella, 1999) and frequently discussed as a corollary of other processes such as globalization (James, 2000), tourism (Scheyvens, 2002) or sustainability and overconsumption (see Cohen and Murphy, 2001; Redclift, 1996) (see Figure 7.1).

> **BOX 7.2 FOOD DESERTS: DEPRIVATION AND DIET**
>
> Neil Wrigley, Daniel Warm and Barrie Margetts' (2003) research on 'Food deserts' is notable in highlighting consumption in terms of material provisioning and issues of need. Addressing social exclusion was a particular priority of the Labour government elected in 1997 in the UK under the leadership of Tony Blair. As part of this, the concept of 'Food deserts' was taken up in policy rhetoric. This term described 'the nutrition and public health problems of those areas of poor retail access' (2003: 153). Wrigley and his colleagues conducted research into the impacts of the opening of a large new Tesco superstore on food consumption patterns in a deprived area of the city of Leeds. As part of the research, seven-day food consumption diaries were filled out by participants and questionnaires were administered both before and after the building of the superstore. The researchers were aware that improved physical access did not imply economic access for participants, and that diet, food purchasing and consumption patterns were influenced by social and cultural norms, facilities for preparation and food practices and other social factors. Nevertheless, the team found there were positive dietary impacts for some of the most vulnerable groups in the area and a modest improvement in diet as a consequence of the retail store intervention in the area.
>
> As well as highlighting an under-researched area within geography, that is the consumption practices of those who may be excluded from participation in retail spaces (for a variety of reasons), Wrigley et al.'s study also points to the ways in which state policy is linked to particular moral discourses of consumption and production (in this case redistributive and integrationist discourses, and the notion of an underclass) and to the power of defining consumption activity (as a food desert) as a means of legitimizing particular policy interventions (2003: 178).

Wrigley et al.'s research on 'food deserts' highlights a dilemma raised by Ramírez (2000): how best to recognize and address geographies of exclusion within consumption research. Ramírez suggests this involves constructing a 'space of discussion and action depending upon and despite our differences' (2000: 540). This remains a challenge and one which involves thinking about the subjects we discuss as well as how we will communicate them – a notion underpinned by moralities, the subject of the next section.

Moral Geographies of Consumption

Discussions about moral imaginings in geography appear to have burgeoned in recent years (Hay, 1998; Proctor and Smith, 1999; Smith, 1997; 1999; 2001). It has been demonstrated elsewhere in this book that consumption geographies are moral constructs. Moralities and moral effects are not limited to certain

FIGURE 7.2 Moralities of consuming. The challenge of developing a critical political edge, involving active practice as well as intellectual transformation (Herb cartoon, permission by Chris Beard)

behaviours (e.g. nineteenth century middle class women who shoplifted) but are also inherent in the perspectives which frame consumption (Miller, 2001c; see Figure 7.2 and Box 7.3).

BOX 7.3 COMMODIFICATION, CONSUMER CULTURE AND MORAL ECONOMY

In a paper in *Environment and Planning D* Andrew Sayer (2003) endeavours to highlight the moral dimensions of consuming and how these moralities have become attached to the ways in which consuming has been understood. Different moralities are constituted by focusing on different aspects of relationships between subjects, objects and space. Sayer suggests that consumption which is beneficial for one relation (such as a parent purchasing for a child) could be damaging for other sorts of social-material relations (such as between rich and poor consumers).

Sayer argues, as does Daniel Miller (2001c), that criticisms of consumption are based on 'elitist prejudices rather than on empirical evidence of how people consume' (2003: 344). Critiques of consumer culture posit an overly negative view of consumption (as narcissistic, superficial, individualistic, destructive) because they view commodification from the point of view of the seller rather than the consumer/user, thus largely ignoring the ways in which commodities are decommoditized and recontextualized or re-enchanted after purchase – practices which are concerned with use and sign value rather than exchange value. Sayer shows that use values tend to be qualitative,

> attached to the character of the object and the subject, while exchange value is quantitative – a determinant of how much one will exchange for another (2003: 345). Because commodification has often been viewed in relation to the sphere of production defined in terms of wage relation and labour markets, it has failed to recognize the qualitative importance of consumption in relation to the goals and perspectives of consumers. Ironically consumption and the commodification of goods offer consumers ways of decommoditizing objects (Kopytoff, 1986). In addition, because such practices are influenced by moral sentiments, norms and prescriptions (such as altruism, showing love or care for others, building relationships etc.), the moral worth and moral behaviour in consumption have also been overlooked.

The insights of new cultural geography and poststructuralism have meant researchers may be more aware of openness, difference and the construction of knowledge claims which lead to particular inclusions and exclusions (Popke, 2003). Place, people and entities exist in relation (the form and nature of these relations differ according to different perspectives), so understanding how relations and places are considered in terms of moralities (good or bad, worthy/unworthy, included and excluded) is a necessary part of understanding how power (both constraining and enabling) and knowledge define particular subjects and objects of consumption (see Box 7.3). It is also important to ask questions about who gets positioned to make assumptions on behalf of 'other groups', in order to develop an understanding of the moral foundations upon which such assumptions are 'built' and the means by which we evaluate them. This implies a need to develop sensitivity not just to others but for others – a sensitivity which has a critical political edge, involving active practice as well as intellectual transformation (Cloke, 2002: 590).

Thus questions about 'the work theoretical perspectives do' imply a need to make both moral framings and consequences viable, despite an inability to ever fully articulate one's partiality or positionality. Ideologies of consumer culture, for example, link hedonism with consumerism and tend to portray consumption as a playground for freedom and the dreamworlds of affluent, narcissistic and self-absorbed consumers (Bauman, 1992). The Marxian notion of the 'commodity fetish' argues commodification, mass culture and materialism provide poor substitutes for 'authentic' meaningful social relations which existed in a previous mode of production when workers were not alienated from the products of their own labour (Wilk, 2001). Veblen's view of consumption was also implicitly negative, with emulation being associated with the wasteful, extravagant and empty patterns of the elite leisure class; while Bourdieu's concept of 'cultural capital' was linked to the accumulations of morally superior knowledges through commodity acquisition.

Moving beyond the work theoretical approaches do to considering what they can do, it is also important to consider the possibilities implied for moral praxis. ANT approaches, for example, imply a relational moral philosophy which shifts attention from ethics emerging primarily from human socialities to those constructed around the relationships between human and non-human actants. ANT approaches may speak to different concerns than perspectives which accord human's primary agency, such as biopolitical activism or animal–human relations (or, as Whatmore, 2003: 139 argues, a case for 'more-than-human geography'). The performance of a commodity chain through consumer activism also re-cites a moral geography, one which asserts labour rights are fundamental human rights (Johns and Vural, 2000). Such performances make some things visible (workplace practices at distant sites, possibilities of change created by consumer activism) and others invisible (potential plant closures and job losses for workers in firms whose products are boycotted). The ways in which consumption practices, events, relationships and spaces are conceptualized may have real impacts for how they are articulated in moral discourses and actualized in time/place contexts. In the United States the assumption that shoplifting in the nineteenth century was a working class rather than a middle class phenomenon (see Box 2.4) demonstrates that as moral framings of consumption practices and behaviours change, so do the appropriate action and response, subject and context. Moral dispositions emplace different entities (such as commodities, subjects, institutions) in social-spatial relationships, which may result in differing inclusions and exclusions and censures. Consider, for example, the emplacement and embodiment of subjects who transgress moral norms in consumption (such as the 'wasteful' consumer) and how these constructions are constituted in relation to and differ across various contexts (for example, in public or private space, or at the level of the individual or the state).

Moral geographies of consumption thus occupy a social and structural position, naming and norming others and contributing to the ways in which the material world is organized and operates. A challenge is to name the sites and subjects of social, cultural, economic and environmental exploitation without doing symbolic injustice to them (Castree, 2001). Mort (1988) cautions that understanding consumption perspectives and practices as moral entities can easily degenerate into a politics of moralism. Criticism of the materialistic and self-indulgent nature of consumption, for example, can result in enunciations of 'the rich' as leading corrupt and superficial lifestyles and 'the poor' as seduced and having material aspirations or consuming beyond their means (Miller, 2001c). An outcome of this can be an ethic which assuages the anxieties of the rich by focusing upon a 'passionate desire to eliminate poverty' (2001c: 227). Miller suggests that humanity needs more, not less, human consumption and that all life involves materialism. Wilk (2001) agrees with Miller (2001c) that we need to carefully analyse the basis of assumptions upon which inevitable moral pronouncements and practices are made and justified (such as caricatures of the poor), but like Sayer (Box 7.3) he argues 'good' moral choices for some may be destructive for others. Wilk seeks to provide a means of engaging 'directly with real fundamental and perennial problems that all human beings have in common in dealing with each other and the material world' (2001: 254–5). Low and Gleeson (1998) suggest that achieving the elimination of poverty, a fair and just distribution of good and bad environments including access to commodities, and abundance and diversity in human and non-human natures and resources, can be seen as reasonable and common goals for humankind. How to intervene to achieve 'just distribution' remains a problem, something Wilk (2002) suggests is compounded by perspectives of consumption which operate as 'islands of practice' (Purcell, 2003), constraining the possibilities for intellectual exchange and limiting possibilities for addressing inequity in consumption (see Box 7.4).

BOX 7.4 PERSPECTIVES AND PRAXIS: INFLUENCING GLOBAL ENVIRONMENTAL CHANGE?

In a paper focusing on 'Consumption, human needs, and global environmental change' Wilk (2002) suggests perspectives on consumption offer different but often unconnected insights into the causes of consumption, the levels and scales at which it is manifest and appropriate policy directions. In considering policies designed to reduce consumption, for example, perspectives which suggests consumption is a response to individual choice and needs driven behaviour would advocate changing the environment in which people make decisions (increasing taxes on vehicles or fuel, more public transportation, more education). When consumption is seen as a social phenomenon concerned with collective behaviour and social distinction, change might be directed at addressing unequal distribution of wealth, with attention given to particular groups and access to forms of consumption. When consumption is viewed as a cultural process, Wilk suggests emphasis on consumption as a meaningful, expressive and symbolic act would imply altering values and beliefs to change consumption practices.

Wilk makes the point that these understandings of consumption and appropriate policy responses tend to ignore the research and insights emanating from the other perspectives. To counter this Wilk (2002: 10) suggests Bourdieu's concept of the habitus (see Box 4.1) – the dispositions, feelings and rules which unconsciously shape and guide behaviour – can enable one to incorporate differing perspectives on how consumption is constituted. A focus on habitus would enable researchers to understand the limits, institutions, impulses, incentives and meanings that guide consumption at diverse levels and scales. Wilk believes the concept is particularly useful for understanding how 'wants' come to be defined as 'needs', emerging in practices which draw

> from one's taken-for-granted habitus. He suggests such an approach would facilitate a more integrated view of the multiple determinants of consumption across a variety of scales, providing some insight into why, for example, most of the negative consequences of consumption appear at the level of regions or nations (as in high levels of greenhouse gas emissions, consumption of fossil fuels etc.), and how these are related to consumption practices of individuals and groups at other scales. Wilk's own perspective is informed by a concern with the significance of (over)consumption in global environmental change. Consequently he views habitus as offering a means to link analysis with praxis and the formulation of policies and strategies which might encourage sustainable consumption and address environmental problems. A potential contribution of the application of habitus is to understand how consumption may be taken for granted, becoming naturalized, in bodily, governmental, social and individual practices – a theme geographers have explored through a number of contexts.

If, as Wilk (2002) suggests, consumption is pivotal in driving environmental change at global and local levels, then knowledge about what determines and changes consumption levels and behaviours is critical. The malnourishment of the urban poor in many developing countries, for example, is related to a growing urban consumer market for food, unevenly developed production and consumption networks at local, regional and global levels, and rapid urbanization (Smith, 1998). Such research need not be informed by the positioning of such consuming subjects as 'victims'. Research on the strategies of the poor in Mexico (Heyman, 2001), on the practices of the homeless in Britain (Cloke, Milbourne and Widdowfield, 2003) and on survival strategies in post-communist households (Smith, 2002) emphasize the agency of those whose consumption choices are constrained. Similarly, Beaverstock et al. (2004: 405) suggest that much can be gained by studying the 'super rich', insights into the ways in which transnationality and global reach might be constructed, and into how consumption practices and spaces link into global flows of labour, commodities and ideas (see for example Lees (2003) on super-gentrification in New York).

Thus understanding how consumption practices create uneven geographies might involve considering how ethical behaviours and subjectivities are framed personally (in the purchase, use and experience of commodities) and collectively (through such things as 'ethical or green consumerism': see Box 1.6) at various levels (the individual, the household, the nation-state) and how these resonate with other knowledges, people and places. It could also involve understanding how discourses of consumption (e.g. as sustainable consumption, or neoliberal consumer sovereignty) operate to define behaviours or actions, and the manifestations of these in place (see Box 7.5).

> **BOX 7.5 QUESTIONS OF SUSTAINABILITY: HOUSEHOLD CONSUMPTION PRACTICES**
>
> Kersty Hobson (2003) argues that a critical geography of consumption can have a significant role in the practice of sustainable consumption – a practice which she argues has been marginalized in high income countries, the very ones with high resource consumption and waste patterns. Agenda 21, 'the blueprint for sustainable development' which emerged from the 1992 UN Conference on Environment and Development in Rio de Janeiro, promotes sustainable consumption as 'The use of goods and services that respond to basic needs and bring a better quality of life, whilst minimizing the use of natural resources, toxic material and emissions of waste and pollutants over the life cycle, so as not to jeopardize the needs of future generations' (IISD/United Nations Department of Economic and Social Affairs, 1999: 1, cited in Hobson, 2003: 148–9).
>
> While Agenda 21 advocates changes at a variety of levels from states to institutions and individual citizens, it has not been considered a pressing environmental concern in Australia. Hobson cites numerous reasons for this: the importance of consumption as part of economic systems predicated on economic growth; the political untenability of governments trying to regulate firm and individual resource use; and the centrality of consumption to the values, lifestyles and everyday practices and habits of individuals. In acknowledging consumption is experienced and practised as part of the often complex, invisible and taken-for-granted operation of households, Hobson suggests that a critical geography of consumption must consider what consumption means for individuals as a practice and a cultural norm. Hobson argues that 'only when we know why and how individuals consume and how they link their consumption to the environment, can we realistically set about changing consumption practices' (2003: 150).
>
> Hobson's politics of consumption is thus different to the one advocated by Hartwick (2000) which involved revealing production and consumption linkages across commodity chains as a basis for effecting 'positive' change. While the two forms are predicated on normative notions of transformation, both argue for a more critical and reflexive engagement with consumption as a meaningful and significant social practice, and as a factor in making and transforming place.

Research is emerging on the institutions and entities through which such discourses circulate, how practices are governed, and the sometimes contradictory spatial and social manifestations of this for consuming in place (see Box 7.6). For example, the construct of the 'international community' in the form of the United Nations is a way of reminding nation-states of the common humanity of their citizens – 'the single strongest slogan of the liberal value of empathy at a distance' – yet ironically it operates as 'a club for the world's most wealthiest nations' (Appadurai, 2002: 43). How institutions and people shape geographies of consumption is therefore a significant issue. Watts and Watts (1983), for example, demonstrated famine in northern Nigeria was not simply a natural phenomenon, but one which was socially constructed, exacerbated by colonial powers who undermined indigenous ways of coping.

BOX 7.6 SHAPING MORAL GEOGRAPHIES:
CONSUMER CITIZENS AND CLIMATE PROTECTION

Rachel Slocum's (2004) research on the Cities for Climate Protection (CCP) campaign highlights the ways in which citizens are shaped as consumers within neoliberal state politics. She examined the implementation of the CCP campaign in Minneapolis, Tuscon and Seattle which had signed on to the worldwide campaign in order to reduce greenhouse gas emissions by changing urban energy use and practices of waste production and transportation. Slocum, drawing on Foucault, notes how the campaign administered through the neoliberal local state produces the subject of the 'consumer citizen'.

Neoliberal discourse situates policy debates within the imperatives of the market, and positions people as consumers and profit maximizers, normalizing particular understandings of value in terms of saving money and cost efficiency. Slocum interviewed representatives of local NGOs, city bureaucrats, active citizens, politicians and business leaders, finding that around 90 per cent of respondents believed climate change should be solved by addressing the 'cost-saving benefit of energy efficiency' (2004: 768). The normalization of this discourse relies on the construction of particular forms of universality: a universal means of valuing the environment, an undifferentiated citizen with universal cost saving concerns, and the closure of alternative options. Thus the CCP campaign reduced climate protection to a commodity (measured in money terms) and offered the solution in terms of a single approach of cost saving, establishing the 'facts' of the argument and ignoring other rationales for and debates about climate change. CCP administrators saw a passive politics relying on consumer education and information (e.g. on energy saving practices) as the way to achieve climate change, thus ignoring the complex reality of the framing of household consumption choices. However, Slocum noted that more progressive possibilities were implied in discourses of consumer citizens shaped by the state (for example, in making governments as deliverers of services more responsive to people). She asserts subjects are able to alter the discourses that constitute them and to actively shape a different or radical politics.

Slocum illustrates how the construction of 'consumer citizens' presents both problems and possibilities for the politics of climate change. In exploring relationships between state and civil society in relation to climate change she believes 'neither a dismissal of consumer politics nor its celebration is warranted' (2004: 767). Her research highlighted the way in which discourses are not ageographical but are produced through practice in particular social-spatial contexts, framed, negotiated and expressed in heterogeneous ways by various agents, implying different consequences and possibilities for change.

The norming and framing of moralities of consumption are also manifest in the growing ethical practices and discourses of consumption, production and trade (Blowfield, 1999; Hale and Shaw, 2001) and the literature connecting consumers with discourses and practices of corporate social responsibility (Maynard, 2001). Interrogating the basis and effects of moral constructions and pronouncements may go some way towards Wilk's (2001) call to link discussions of consumption with fundamental problems of justice and the common good. How geographers address and perform such responsibilities towards others remains a challenge, not least because of an inability to know the multiple places we speak from and to, or the complex ways in which our own moralities are (re)produced. This need not, however, preclude efforts to invoke positive change.

Politics and praxis

How to critically engage with different moral positions and to create a space of deliberation with transformative potential is something geographers have begun to consider with regard to consumption. In *Spaces of hope* (2000), David Harvey endeavours to frame an alternative social project based on utopian imagining of a more socially just society. Although he endeavours to link economic and cultural power (for example through his work on the body and globalization, see also (1998)), his perspective situates the body as a site of emancipatory politics within the logic of capital, limiting the agency of the pleasurable, the felt and expressed body. Sack's (1997; 1999) use of his relational framework (see Box 3.2) concerning meaning, nature and social relations to develop a geographical theory of morality is also located within the sphere of production, based on the immorality of the fetishization of commodity which presents consumption as an act with no moral consequences, disguising the (often exploitative) relationships between production, people and nature that make production and consumption of commodities possible. Though he privileges conceptions of a universal global morality over local and context based moralities, Sack and other geographers who have sought to explore cultural politics explicitly highlight the ways in which morality and power might be connected to position people as subjects. The challenge remains of how to engage in transformative possibilities in ways which are not deterministic, which recognize the inseparability of consumptions and production, yet provide a space and a politics which enable real change to occur.

Addressing issues of social justice in consumption geographies must also involve a consideration of the 'power geometries' and moralities which result from geographers' practice, teaching and research. This could involve considering how some perspectives on consumption circulate as universal or mainstream while others are viewed as social or peripheral, and how 'successful' performances of knowledge might deny, limit or make invisible other research performances. Examining how power and knowledge are connected, circulated and constructed through production and consumption relationships is important to establishing a politics of action (Bell and Valentine, 1997). Though such knowledge will always be partial, critical geographies of consumption should endeavour to foreground the fabrications made available at different sites (Crang, 1996).

Notwithstanding the difficulty of engaging in research with participants for whom consumption choices might be severely limited, largely unconstrained, or simply very different from one's own, engaging with the narratives of 'distant' others is important. A commitment to others and a sense for others could mean, for example, understanding how orientation as consuming subjects may be experienced as marginalization or exclusion in places geographically proximate or physically distant.

So how might recognition of the moral geographies and commitment to others be practised, and how might this inform a grounded moral praxis for us as geographers? Perhaps a key is the development of spaces in which a politics of imagining is connected with the politics of practice. Distancing militates against empathizing with others who may be different (Smith, 1999). A Guide meeting in Mexico in 1981 (discussed in Box 7.7) used a myriad of exercises to explicate and perform difference between self and others in order to motivate and to develop not just empathy but a responsibility toward others.

BOX 7.7 THE PRACTICE OF HUNGER

In New Zealand, advertisements for child sponsorship, and print and television reports of famine and war, bring the stories of distant 'others' into many living rooms. Representations of daily non-consumption or underconsumption from organizations such as Greenpeace, Oxfam and World Vision frame space in terms of a production/consumption gap, typically characterized by statistics which exemplify and construct First World and Third World as rich and poor. Citations of statistics such as 'the net wealth of the ten richest billionaires is $133 billion, more than 1.5 times the total nation income of the least developed countries' (United Nations Development Programme, 1999) and '20 per cent of the world's population is responsible for 80 per cent of greenhouse gas emissions' (Wilk, 2001: 356) reify the notion of a consumption/production gap between First and Third world despite the fact that this gap increasingly exists within nations of the First World too. Access to, choice of and freedom to consume and use commodities creatively in consumption practices are unevenly developed. Miles, for example, illustrates the contradiction inherent in structures of capitalism when he notes 'the system of choice that constitutes the "free" individual in the West also generates massive oppression in as much as those excluded from making such choices become disenfranchised and oppressed' (1998a: 150).

While media and popular representations appear to bring people closer to the lived experience, livelihood and life chance dilemmas that are a feature of the bulk of the world's population, they do not simultaneously confront the spaces of politics that are integral to consumption (Le Heron, 2003).

A 10-day international Girl Guide leaders' meeting in Mexico in 1981 endeavoured to confront such a politics of space, through creating a 'space of politics' aimed at encouraging participants to engage with issues of environment, hunger and the status of women.[5] In bringing together 30 leaders from 18 countries (of varying 'developmental' status), the meeting was intended to make consumption dilemmas and differences real for the participants.[6] The intention of the meeting was to encourage commitment to 'global' issues and to stimulate practical action in leaders' home countries on their return.

As part of a 'Bread Alert' exercise (see Figure 7.3) the leaders were randomly divided into groups representing the approximate proportions of First World, Second World and Third World at that time (in a ratio of 3:7:20 respectively) and given breakfast: First World, a full 'American' style breakfast; Second World, a small glass of fruit juice and a roll; Third World, a dessertspoon of boiled rice. Barbara Arnold, one of the two New Zealand delegates, was assigned to the First World group. She tells how difficult it was to sit next to others who had little or nothing, knowing that for some of the people in these groups such consumption levels might be a daily reality. While Barbara and her First World colleagues wanted to cooperate to give the others a portion of their allowance, they were forbidden to. Barbara noted the situation was made worse by the others having to watch her group eat, and she wonders whether they would have felt the same if the Second and Third World groups had not remained at the breakfast table. Thus the realities of the participants' consumption, and spaces of politics which separated them from understanding and acting toward others, were made visible in the performance of hunger in a context in which a visual and embodied exchange could take place. As part of the meeting participants were required to undertake other Bread Alert exercises aimed to confront passivity through a revolution 'that begins in the stomach'.

If the central issue in social justice is that of identifying those differences among persons and groups which are morally significant to the distribution of benefits and burdens (D.M. Smith, 2000: 1150), then the Bread Alert exercise provides a forum for such recognitions upon which forms of political action might be undertaken. The meeting suggested a number of possible post-meeting activities as a way of making a difference: disseminating information in public forums, influencing Guides' understanding of such issues, changing purchase and disposal practices, organizing campaigns, holding foodless banquets, and lobbying decision makers.

The Bread Alert exercise also demonstrates how spaces of politics may be confronted by a situated politics of space. By being put into situations which may or may not reflect the consumption decisions of their normal everyday lives, participants were made aware of the way in which geographies (in this case constituting hunger) were embodied and emplaced. Barbara saw how food consumption was embedded in practices which are performed, travelled and translated to produce particular outcomes and

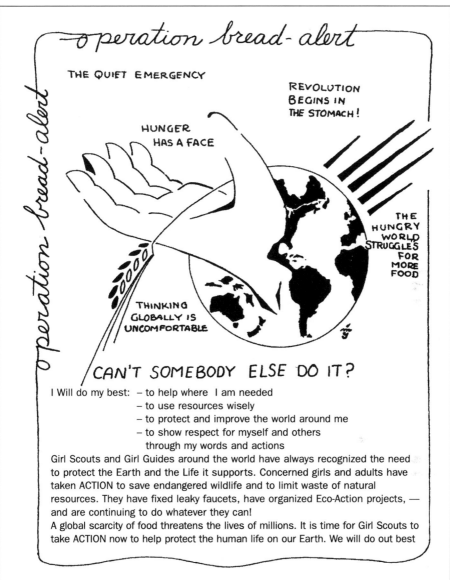

FIGURE 7.3   Practising hunger: Operation Bread Alert, Mexico, 1981 (permission by *Girl Guides, New Zealand*)

connections, the politics of which may not easily be altered (as Barbara found when she wanted to coopt to share). Consumption geographies are therefore geographies of embodiment, embeddedness, performance and travel. The Mexico case study demonstrates the interconnections between the performance of spatialities (uneven development re-citing differences in food choices) and the spatialities of performance (as

> groups were obliged to perform competition through assigned roles in a particular context). The frustrations that Barbara and her colleagues experienced demonstrated how the connections/journeys between self and others are created not only through material practice, but also through spatial/social imaginings.

When consumption is conceived as a geography of embodiment, embeddedness, performance and travel, it does not restrict consumption to particular consumption sites and spaces, to the end of a commodity chain, to a circuit of culture, to practices of self-identification or subjectification, to the symbolic and material appropriation of commodities in everyday life, or to following the biographies, histories and geographies of commodities. Rather knowledges about consumption can be seen as performative of these things, circulating and being invested with power in situated contexts and providing different insights on the world. Under such a schema, consumption does not come to mean everything and nothing, but takes its form and is actualized in the conditions of its reproduction. It is crucial to explore the (moral) spaces of visualization, embodiment and translation that result from the seeing, doing and becoming of consumption as a situated social practice, and as part of the practice of geography. Doing so may assist in understanding and addressing the undesirable and uneven consequences of the power geometries which result.

Consuming Geographies

A view of consumption geographies as performative, as situated geographies which are re-cited, travelled and transformed, forming different power geometries, suggests that conceptualizing and researching consumption are not simply constituted in representation, meaning and identity but are simultaneously grounded in non-representational processes, practices and structures in place. When the focus is on the work that perspectives, people and entities do, their associations, moralities, possibilities and limitations, and how they flow and are translated in place, the binaries of production/consumption, culture/economy, material/symbolic no longer become central. Instead the emphasis is on the situated articulation of consumption knowledges, practices and relations and how these bring different subjects and spaces into focus and evoke different kinds of politics, moralities and possibilities for action.

This textbook has focused on how consumption matters to socialities, subjectivities and spatialities as they are reproduced and reflected in processes of place, scale and boundary making. It is my hope that readers have caught a glimpse of the significant and fascinating insights consumption geographies have provided into these things as they are constructed through place, space and scale and across boundaries. It is also my hope that this text will encourage readers to ask questions – not about 'whether consumption matters', but about how and why consumption is so central to 'geography matters'!

FURTHER READING

Cloke, P. (2002) 'Deliver us from evil? Prospects for living ethically and acting politically in human geography', *Progress in Human Geography*, 26 (5): 587–604.

Cohen, M.J. and Murphy, J. (eds) (2001) *Exploring Sustainable Consumption: Environmental Policy and the Social Sciences*. Oxford: Pergamon. pp. 121–33.

Miller, D. (2001) 'The poverty of morality', *Journal of Consumer Culture*, 1 (2): 225–43.

Purcell, M. (2003) 'Islands of practice and the Marston/Brenner debate: toward a more synthetic critical human geography', *Progress in Human Geography*, 27 (3): 317–32.

Redclift, M. (1996) *Wasted: Counting the Costs of Global Consumption*. London: Earthscan.

Sack, R.D. (1997) *Homo Geographicus: a Framework for Action, Awareness, and Moral Concern*. Baltimore, MD: Johns Hopkins University Press.

Wilk, R. (2001) 'Consuming morality', *Journal of Consumer Culture*, 1 (2): 245–60.

NOTES

1. That knowledges are performative is implicit in the writings of Lakoff and Johnson (1980) and Olsson (1980) on the theory dependence of observation.
2. I am indebted to Richard Le Heron and Wendy Larner (University of Auckland) for their comments on my ideas for this chapter. The two questions 'What work do they do?' and 'What work can they do?' are attributed to Wendy, and form an important part of Larner and Le Heron's emerging work on poststructural political economies (Larner and Le Heron, 2002a; Greenaway et al., 2002).
3. Crang's (1997) work on spaces of picturing in tourism, and Desforges' (2001) work on the construction of money in tourism, have encouraged me to think about whether the 'practices of picturing' through different geographical perspectives on consumption can open up possibilities for considering how the subjects and objects of consumption are embedded in time and place (Crang, 1997: 331).
4. Anglo-American geographies should nevertheless be considered a hybrid construction which does draw on non-Anglo-American discourses (Samers and Sidaway, 2000).
5. My sincere thanks to Barbara Arnold for providing me with this information, and for taking the time to tell me about her experience.
6. Numbers from each country: Australia (2), UK (2), Bahamas (2), Zimbabwe (1), Liechtenstein (1), Norway (2), Peru (2), Honduras (1), Canada (2), USA (2), Jamaica (2), Japan (2), Mexico (2), New Zealand (2), Netherlands Antilles (2), Trinidad and Tobago (1), Finland (1), Guatemala (2), Panama (1).

References

Abaza, M. (2001) 'Shopping malls, consumer culture and the reshaping of public space in Egypt', *Theory, Culture and Society*, 18 (5): 97–122.

Abelson, E. (2000) 'Shoplifting ladies', in J. Scanlon (ed.), *The Gender and Consumer Culture Reader*. New York: New York University Press. pp. 309–29.

Ackerman, F. (1997) 'The history of consumer society: overview essay', in N.R. Goodwin, F. Ackerman and D. Kiron (eds), *The Consumer Society: Frontier Issues in Economic Thought*, vol. 2. Washington, DC: Island. pp. 109–18.

Adorno, T. and Horkheimer, M. (1944) *Dialectic of Enlightenment*. London: Verso.

Aitchison, C. (1999) 'New cultural geographies: the spatiality of leisure, gender and sexuality', *Leisure Studies*, 18 (1): 19–39.

Aitchison, C., MacLeod, N.E. and Shaw, S.J. (2000) *Leisure and Tourism Landscapes: Social and Cultural Geographies*. London: Routledge.

Aldridge, T.J. and Patterson, A. (2002) 'LETS get real: constraints on the development of Local Exchange Trading Schemes', *Area*, 34 (4): 370–81.

Amin, A. (2002) 'Spatialities of globalisation', *Environment and Planning A*, 34 (3): 385–99.

Appadurai, A. (ed.) (1986) *The Social Life of Things: Commodities in Cultural Perspective*. Cambridge: Cambridge University Press.

Appadurai, A. (2002) 'Broken promises', *Foreign Policy*, 132 (Sep–Oct): 42–44.

Argenbright, R. (1999) 'Remaking Moscow: new places, new selves', *Geographical Review*, 89 (1): 1–22.

Ateljevic, I. (2000) 'Circuits of tourism: stepping beyond the "production/consumption" dichotomy', *Tourism Geographies*, 2 (4): 369–88.

Azaryahu, M. (1999) 'McDonald's or Golani Junction? A case of a contested place in Israel', *Professional Geographer*, 51 (4): 481–92.

Backes, N. (1997) 'Reading the shopping mall city', *Journal of Popular Culture*, 31 (3): 1–17.

Baker, P. (2001) 'The new Moscow', *Sunday Star Times*, p. C16.

Bakhtin, M. (1984) *Rabelais and His World*. Bloomington, IN: Indiana University Press.

Banim, M., Green, E. and Guy, A. (2001) 'Introduction', in A. Guy, E. Green and M. Banim (eds), *Through the Wardrobe: Women's Relationships with their Clothes*. Oxford: Berg. pp. 1–17.

Barber, K. and Waterman, C. (1995) 'Traversing the global and the local: *fuji* music and praise poetry in the production of contemporary Yoruba popular culture', in D. Miller (ed.), *Worlds Apart: Modernity through the Prism of the Local*. London: Routledge. pp. 240–62.

Barker, A.M. (1999) 'Going to the dogs: pet life in the new Russia', in *Consuming Russia: Popular Culture, Sex, and Society since Gorbachev*. Durham, NC: Duke University Press. pp. 12–45.

Barnett, J.R. and Kearns, R.A. (1996) 'Shopping around? Consumerism and the use of private accident and medical clinics in Auckland, New Zealand', *Environment and Planning A*, 28 (6): 1053–75.

Barnett, S. (1997) 'Research note. Maori tourism', *Tourism Management,* 18 (7): 471–3.
Bater, J., Degtyarev, A. and Amelin, V. (1995) 'Politics in Moscow: local issues, areas and governance', *Political Geography,* 14 (8): 665–87.
Baudrillard, J. (1981) *For a Critique of the Political Economy of the Sign.* St Louis, MO: Telos.
Baudrillard, J. (1983) *Simulations.* New York: Semiotext(e).
Baudrillard, J. (1988) *Jean Baudrillard: Selected Writings.* Standford, CA: Standford University Press.
Bauman, Z. (1990) *Thinking Sociologically: an Introduction for Everyone.* Oxford: Blackwell.
Bauman, Z. (1992) *Intimations of Postmodernity.* London: Routledge.
Beaverstock, J.V., Hubbard, P. and Short, J.R. (2004) 'Getting away with it? Exposing the geographies of the super-rich', *Geoforum,* 35: 401–7.
Beck, U. (1992) *Risk Society: Towards a New Modernity.* London: Sage.
Beder, S. (2002) 'Nike greenwash over sweatshop labour'. At http://www.theecologist.org/archive_article.html?article=298&category=66, 22 August 2003.
Bell, D. and Valentine, G. (1997) *Consuming Geographies: We Are Where We Eat.* London: Routledge.
Benjamin, W. (1970) *Illuminations.* London: Fontana.
Benjamin, W. (1983) *The Flâneur, Charles Baudelaire: a Lyric Poet in the Era of High Capitalism.* London: Verso.
Bennett, A. (1999) 'Rappin' on the Tyne: white hip-hop culture in Northeast England – an ethnographic study', *The Sociological Review,* 47 (1): 1–24.
Benson, S. (1997) 'The body, health and eating disorders', in K. Woodward (ed.), *Identity and Difference.* London: Sage. pp. 121–81.
Berg, L.D. (1994) 'Masculinity, place and a binary discourse of theory and empirical investigation in the human geography of Aotearoa/New Zealand', *Gender, Place and Culture,* 1 (2): 245–60.
Berg, L.D. and Kearns, R.A. (1996) 'Naming as norming? "Race", gender and the identity politics of naming places in Aotearoa/New Zealand', *Environment and Planning D: Society and Space,* 14 (1): 99–122.
Berg, L.D. and Kearns, R.A. (1998) 'America unlimited', *Environment and Planning D: Society and Space,* 16: 128–32.
Berg, L.D. and Mansvelt, J.R. (2000) 'Writing in, speaking out: communicating qualitative research findings', in I. Hay (ed.), *Qualitative Research Methods in Human Geography.* Melbourne: Oxford University Press.
Berg, L.D. and Roche, M.M. (1997) 'Market metaphors, neo-liberalism and the construction of academic landscapes in Aotearoa/New Zealand', *Journal of Geography in Higher Education,* 21 (2): 147–61.
Bhabha, H.K. (1994) *The Location of Culture.* London: Routledge.
Bhabha, H.K. (2001) 'Locations of culture: the postcolonial and the postmodern', in S. Malpas (ed.), *Postmodern Debates.* New York: Palgrave. pp. 36–44.
Binnie, J. (1995) 'Trading places: consumption, sexuality and the production of queer space', in G. Valentine and D. Bell (eds), *Mapping Desire: Geographies of Sexualities.* London: Routledge. pp. 182–99.
Binyon, M. (2002a) 'Booming Russia has happy Christmas', *The Dominion,* 8 January: 8.
Binyon, M. (2002b) 'Moscow considers curfew for street kids', *The Dominion,* 26 January: 28.
Blair, J. and Gereffi, G. (2001) 'Local clusters in global chains: the causes and consequences of export dynamism in Torreon's blue jeans industry', *World Development,* 29 (11): 1885–903.

Blomley, N. (1996) '"I'd like to dress her all over": masculinity, power and retail space', in N. Wrigley and M. Lowe (eds), *Retailing Consumption and Capital*. Harlow: Longman. pp. 235–56.

Blowfield, M. (1999) 'Ethical trade: a review of developments and issues', *Third World Quarterly*, 20 (4): 753–70.

Bocock, R. (1993) *Consumption*. London: Routledge.

Bourdieu, P. (1984) *Distinction: a Social Critique of the Judgement of Taste*, trans. R. Nice. Cambridge, MA: Harvard University Press.

Bowlby, R. (1985) *Just Looking: Consumer Culture in Dreiser, Gissing and Zola*. New York: Methuen.

Bowlby, S., Gregory, S. and McKie, L. (1997) '"Doing home": patriarchy, caring, and space', *Women's Studies International Forum*, 20 (3): 343–50.

Bowler, S.M. (1995) 'Managing the shopping centre as a consumption site. Creating appealing environments for visitors: some Australian and New Zealand examples'. Unpublished PhD thesis, Massey University, Palmerston North, New Zealand.

Bridge, G. and Dowling, R. (2001) 'Microgeographies of retailing and gentrification', *Australian Geographer*, 32 (1): 93–107.

Bridge, G. and Smith, A. (2003) 'Guest editorial. Intimate encounters: culture – economy – commodity', *Environment and Planning D: Society and Space*, 21: 257–68.

Bryman, A. (1999) 'The Disneyization of society', *The Sociological Review*, 47 (1): 25–47.

Buck-Morss, S. (1989) *The Dialectics of Seeing: Walter Benjamin and the Arcades Project*. Cambridge, MA: MIT Press.

Burrows, R. and Marsh, C. (eds) (1992) *Consumption and Class*. London: Macmillan.

Burt, S. and Sparks, L. (2001) 'The implications of Wal-Mart's takeover of Asda', *Environment and Planning A*, 33 (8): 1463–87.

Butler, J. (1990) *Gender Trouble: Feminism and the Subversion of Identity*. New York: Routledge.

Butler, R.W. (1991) 'West Edmonton Mall as a tourist attraction', *The Canadian Geographer*, 35 (3): 287–95.

Cahill, S. and Riley, S. (2001) 'Resistances and reconciliations: women and body art', in A. Guy, E. Green and M. Banim (eds), *Through the Wardrobe: Women's Relationships with their Clothes*. Oxford: Berg. pp. 151–70.

Callard, F.J. (1998) 'The body in theory', *Environment and Planning D: Society and Space*, 16: 387–400.

Cameron, J. (1998) 'The practice of politics: transforming subjectivities in the domestic domain and the public sphere', *Australian Geographer*, 29 (3): 293–307.

Campbell, C. (1987) *The Romantic Ethic and the Spirit of Modern Consumerism*. Oxford: Blackwell.

Campbell, C. (1995) 'Conspicuous confusion? A critique of Veblen's theory of conspicuous consumption', *Sociological Theory*, 13: 37–47.

Campbell, H. and Liepins, R. (2001) 'Naming organics: understanding organic standards in New Zealand as a discursive field', *Sociologia Ruralis*, 41 (1): 21–39.

Carroll, J. and Connell, J. (2000) '"You gotta love this city": The Whitlams and inner Sydney', *Australian Geographer*, 31 (2): 141–54.

Castells, M. (1996) *The Rise of the Network Society*, vol. I. Cambridge, MA: Blackwell.

Castree, N. (2001) 'Commentary. Commodity fetishism, geographical imaginations and imaginative geographies', *Environment and Planning A*, 33: 1519–25.

Chandran, R. (2002) 'Trash e-trash'. At http://www.blonnet.com/ew/2002/04/10/stories/2002041000130400.htm, 3 May 2004.

Chatterton, P. and Hollands, R. (2002) 'Theorising urban playscapes: producing, regulating and consuming youthful nightlife city spaces', *Urban Studies*, 39 (1): 95–116.

Clancy, M. (1998) 'Commodity chains, services and development: theory and preliminary evidence from the tourism industry', *Review of International Political Economy*, 5 (1): 122–48.

Clancy, M. (2002) 'The globalization of sex tourism and Cuba: a commodity chains approach', *Studies in Comparative International Development*, 36 (4): 63–88.

Clarke, A.J. (1998) 'Window shopping at home: classifieds, catalogues and new consumer skills', in D. Miller (ed.), *Material Cultures: Why Some Things Matter*. London: UCL Press. pp. 73–99.

Clarke, A.J. (2000) '"Mother swapping": the trafficking of nearly new children's wear', in P. Jackson, M. Lowe, D. Miller and F. Mort (eds), *Commercial Cultures*. Oxford: Berg. pp. 85–100.

Clarke, A.J. (2001) 'The aesthetics of social aspiration', in D. Miller (ed.), *Home Possessions: Material Culture behind Closed Doors*. Oxford: Berg. pp. 23–67.

Clarke, B. (1999) '100% pure NZ tourism fake', *Sunday Star Times*, 29 August, p. 3.

Clarke, D.B. and Bradford, M.G. (1998) 'Public and private consumption and the city', *Urban Studies*, 35 (5/6): 865–88.

Clarke, D.B., Doel, M.A. and Housiaux, K.M.L. (2003) 'Introduction to Part Two: Geography', in D.B. Clarke, M.A. Doel and K.M.L. Housiaux (eds), *The Consumption Reader*. London: Routledge. pp. 80–6.

Clarke, J. and Purvis, M. (1994) 'Dialectics, difference, and the geographies of consumption', *Environment and Planning A*, 26 (7): 1091–109.

Cloke, P. (1993) 'The countryside as commodity: new rural spaces for leisure', in S. Glyptis (ed.), *Leisure and the Environment*. London: Belhaven. pp. 53–66.

Cloke, P. (2002) 'Deliver us from evil? Prospects for living ethically and acting politically in human geography', *Progress in Human Geography*, 26 (5): 587–604.

Cloke, P. and Widdonfield, R.C. (2000) 'The hidden and emerging spaces of rural homelessness', *Environment and Planning A*, 32 (1): 77–90.

Cloke, P., Milbourne, P. and Widdowfield, R. (2003) 'The complex mobilities of homeless people in rural England', *Geoforum*, 34: 21–35.

Cloke, P.J., Philo, C. and Sadler, D. (1991) *Approaching Human Geography: an Introduction to Contemporary Theoretical Debate*. London: Chapman.

Cockburn, C. (1997) 'Domestic technologies: Cinderella and the engineers', *Women's Studies International Forum*, 20 (3): 361–71.

Cohen, M.J. and Murphy, J. (2001) *Exploring Sustainable Consumption: Environmental Policy and the Social Sciences*. Oxford: Pergamon.

Connell, J. and Gibson, C. (2003) *Sound Tracks: Popular Music, Identity and Place*. London: Routledge.

Connor, T. and Atkinson, J. (1996) 'Sweating for Nike: a report on labor conditions in the sport shoe industry'. Community Aid Abroad Briefing Paper no. 16. At http://www.caa.org.au/campaigns/nike/sweating.html, 6 August 1998.

Cook, I. and Crang, P. (1996) 'The world on a plate: culinary culture, displacement and geographical knowledges', *Journal of Material Culture*, 1 (2): 131–53.

Cook, I., Crang, P. and Thorpe, M. (1999) 'Eating into Britishness: multicultural imaginaries and the identity politics of food', in S. Roseneil and J. Seymour (eds), *Practising Identities, Power and Resistance*. Basingstoke: Macmillan. pp. 223–48.

Corrigan, P. (1997) *The Sociology of Consumption*. London: Sage.

Crang, M. (1997) 'Picturing practices: research through the tourist gaze', *Progress in Human Geography*, 21 (3): 359–73.

Crang, M. (2002) 'Commentary. Between places: producing hubs, flows, and networks', *Environment and Planning A*, 34: 569–74.

Crang, M., Crang, P. and May, J. (eds) (1999) *Virtual Geographies: Bodies, Space and Relations*. London: Routledge.

Crang, P. (1994) 'It's showtime: on the workplace geographies of display in a restaurant in South East England', *Environment and Planning D: Society and Space*, 12: 675–704.

Crang, P. (1996) 'Displacement, consumption and identity', *Environment and Planning A*, 28 (1): 47–68.

Crang, P. and Cook, I. (1996) 'The world on a plate: culinary knowledge, displacement and geographical knowledge', *Journal of Material Culture*, 1: 131–53.

Crang, P., Dwyer, C. and Jackson, P. (2003) 'Transnationalism and the spaces of commodity culture', *Progress in Human Geography*, 27 (4): 438–56.

Cressy, D. (1993) 'Literacy in context: meaning and measurement in early modern England', in J. Brewer and R. Porter (eds), *Consumption and the World of Goods*. London: Routledge. pp. 305–19.

Crewe, L. (2000) 'Progress reports. Geographies of retailing and consumption', *Progress in Human Geography*, 24 (2): 275–91.

Crewe, L. (2001) 'Progress reports. The besieged body: geographies of retailing and consumption', *Progress in Human Geography*, 24 (4): 629–41.

Crewe, L. (2003) 'Progress reports geographies of retailing and consumption: markets in motion', *Progress in Human Geography*, 27 (3): 352–62.

Crewe, L. and Gregson, N. (1998) 'Tales of the unexpected: exploring car boot sales as marginal spaces of contemporary consumption', *Transactions of the Institute of British Geographers*, NS 23 (1): 39–53.

Crewe, L. and Lowe, M. (1995) 'Gap on the map? Towards a geography of consumption and identity', *Environment and Planning A*, 27 (12): 1877–98.

Crompton, R. (1996) 'Consumption and class analysis', in S. Edgell, K. Hetherington and A. Warde (eds), *Consumption Matters*. Oxford: Blackwell. pp. 113–32.

Cross, G. (1993) *Time and Money: the Making of Consumer Culture*. London: Routledge.

Crouch, D. (ed.) (1999) *Leisure/Tourism Geographies*. London: Routledge.

Dean, M. (1999) *Governmentality: Power and Rule in Modern Society*. London: Sage.

Debord, G. (1994) *Translation: the Society of the Spectacle*. New York: Zone.

De Certeau, M. (1984) *The Practice of Everyday Life*. Berkeley, CA: University of California Press.

DeNora, T. and Belcher, S. (2000) '"When you're trying something on you picture yourself in a place where they are playing this kind of music": musically sponsored agency in the British clothing retail sector', *The Sociological Review*, 48 (1): 80–107.

Desforges, L. (2001) 'Tourism consumption and the imagination of money', *Transactions of the Institute of British Geographers*, NS 26 (3): 353–64.

Desforges, L. and Jones, R. (2001) 'Geographies of languages/languages of geography', *Social & Cultural Geography*, 2 (3): 261–4.

Dionne, E.J. (1998) 'Bad for business', *Washington Post*, 15 May: A27.

Dodge, M. and Kitchin, R. (2000) *Mapping Cyberspace*. New York: Routledge.

Doel, M.A. and Clarke, D.B. (1999) 'Dark panopticon. Or, attack of the killer tomatoes', *Environmental and Planning D: Society and Space*, 17 (4): 427–50.

Domosh, M. (2001) 'The "Women of New York": a fashionable moral geography', *Environment and Planning D: Society and Space*, 19: 573–92.

Domosh, M. (2003) 'Pickles and purity: discourses of food, empire and work in turn-of-the-century USA', *Social & Cultural Geography*, 4 (1): 7–26.

Donaghu, M.T. and Barff, R. (1990) 'Nike just did it: international subcontracting and flexibility in athletic footwear production', *Regional Studies*, 24 (6): 537–52.

Douglas, S.J. (2000) 'Narcissism as liberation', in J. Scanlon (ed.), *The Gender and Consumer Culture Reader*. New York: New York University Press. pp. 267–82.

Douglas, M. and Isherwood, B. (1978) *The World of Goods: Towards an Anthropology of Consumption*. London: Allen Lane.

Dowling, R. (1993) 'Femininity, place and commodities: a retail case study', *Antipode*, 25 (4): 295–319.

Du Gay, P. and Negus, K. (1994) 'The changing sites of sound: music retailing and the composition of consumers', *Media, Culture & Society*, 16: 395–413.

Du Gay, P. and Pryke, M. (2002) 'Introduction', in P. du Gay and M. Pryke (eds), *Cultural Economy: Cultural Analysis and Commercial Life*. London: Sage. pp. 1–19.

Du Gay, P., Hall, S., Janes, L., Mackay, H. and Negus, K. (1997) *Doing Cultural Studies: the Story of the Sony Walkman*. London: Sage.

Dwyer, C. and Jackson, P. (2003) 'Commodifying difference: selling EASTern fashion', *Environment and Planning D: Society and Space*, 21: 269–91.

Edgell, S., Hetherington, K. and Warde, A. (eds) (1996) *Consumption Matters*. Oxford: Blackwell.

Edwards, T. (2000) *Contradictions of Consumption: Concepts, Practices, and Politics in Consumer Society*. Philadelphia, PA: Open University Press.

Entwistle, J. (2000) 'Fashioning the career woman: power dressing as a strategy of consumption', in M. Andrews and M.M. Talbot (eds), *All the World and Her Husband*. London: Cassell. pp. 224–38.

Erkip, F. (2003) 'The shopping mall as an emergent public space in Turkey', *Environment and Planning A*, 35 (6): 1073–93.

Espiner, C. (1999) 'Tourism Board website censured', *The Press*, 23 September: 1.

Espiner, C. (2001) 'Maori seek bigger share of tourism', *The Press*, 21 August: 2.

Falk, P. and Campbell, C. (eds) (1997) *The Shopping Experience*. London: Sage.

Featherstone, M. (1987) 'Lifestyle and consumer culture', *Theory, Culture and Society*, 4 (1): 55–70.

Featherstone, M. (1991) *Consumer Culture and Postmodernism*. London: Sage.

Fine, B. (1993) 'Modernity, urbanism, and modern consumption: a comment', *Environment and Planning D: Society and Space*, 11 (5): 599–601.

Fine, B. (2002) *The World of Consumption: the Material and Cultural Revisited*, 2nd edn. London: Routledge.

Fine, B. and Leopold, E. (1993) *The World of Consumption*. London: Routledge.

Fiske, J. (1989) 'Shopping for pleasure: malls, power, and resistance', in *Reading the Popular*. London: Unwin Hyman. pp. 13–42.

Foucault, M. (1979) *Discipline and Punish: the Birth of the Prison*. New York: Vintage.

Friedman, J. (ed.) (1994) *Consumption and Identity*. Chur, Switzerland: Harwood.

Frith, S. (1996) *Performing Rites: on the Values of Popular Music*. Cambridge, MA: Harvard University Press.

Gamman, L. (2000) 'Visual seduction and perverse compliance: reviewing food fantasies, large appetites and "grotesque" bodies', in S. Bruzzi and P.C. Gibson (eds), *Fashion Culture: Theories, Explorations and Analysis*. London: Routledge. pp. 61–78.

Gamman, L. and Makinen, M. (1994) *Female Fetishism: a New Look*. London: Lawrence & Wishart.

Garcia-Ramon, M.-D. (2003) 'Globalization and international geography: the questions of languages and scholarly traditions', *Progress in Human Geography*, 27 (1): 1–5.

Gelber, S.M. (2000) 'Do-it-yourself: constructing, repairing, and maintaining domestic masculinity', in J. Scanlon (ed.), *The Gender and Consumer Culture Reader*. New York: New York University Press. pp. 70–93.

Gereffi, G. (1999) 'International trade and industrial upgrading in the apparel commodity chain', *Journal of International Economics*, 48: 37–70.

Gereffi, G. (2001) 'Beyond the producer-driven/buyer-driven dichotomy', *IDS Bulletin*, 32 (3): 30.

Gereffi, G. and Korzeniewicz, M. (eds) (1994) *Commodity Chains and Global Capitalism*. Westport, CT: Greenwood.

Gershuny, J. and Miles, S. (1983) *The New Service Economy: the Transformation of Employment in Industrial Societies*. London: Pinter.

Gibson, P.C. (2000) '"No-one expects me anywhere": invisible women, ageing and the fashion industry', in S. Bruzzi and P.C. Gibson (eds), *Fashion Culture: Theories, Explorations and Analysis*. London: Routledge. pp. 79–89.

Gilroy, P. (1993) 'Mixing it – how is British national identity defined and how do race and nation intersect?', *Sight and Sound*, 3 (9): 24–5.

Glennie, P. (1995) 'Consumption within historical studies', in D. Miller (ed.), *Acknowledging Consumption: a Review of New Studies*. London: Routledge. pp. 164–203.

Glennie, P.D. and Thrift, N.J. (1992) 'Modernity, urbanism, and modern consumption', *Environment and Planning D: Society and Space*, 10 (4): 423–43.

Glennie, P.D. and Thrift, N.J. (1993) 'Modern consumption: theorising commodities and consumers', *Environment and Planning D: Society and Space*, 11 (5): 603–6.

Glennie, P. and Thrift, N. (1996) 'Consumption, shopping and gender', in N. Wrigley and M. Lowe (eds), *Retailing, Consumption and Capital: Towards the New Retail Geography*. Harlow: Longman. pp. 221–37.

Goffman, E. (1971 [1959]) *The Presentation of Self in Everyday Life*. Harmondsworth: Penguin.

Goldman, R. and Papson, S. (1998) *Nike Culture: the Sign of the Swoosh*. London: Sage.

Goodall, P. (1991) 'Design and gender: where is the heart of the home?', *Built Environment*, 16 (4): 269–78.

Goodman, D. (1999) 'Agro-food studies in the "age of ecology": nature, corporeality, biopolitics', *European Society for Rural Sociology*, 39 (1): 17–38.

Goodman, D. (2001) 'Ontology matters: the relational materiality of nature and agro-food studies', *Sociologia Ruralis*, 41 (2): 182–200.

Goodman, D. and Dupuis, E.M. (2002) 'Knowing food and growing food: beyond the production–consumption debate in the sociology of agriculture', *Sociologia Ruralis*, 42: 5–22.

Goss, J. (1993) 'The "magic of the mall": an analysis of form, function, and meaning in the contemporary retail built environment', *Annals of the Association of American Geographers*, 83 (1): 18–47.

Goss, J. (1999a) 'Consumption', in P. Cloke, P. Crang and M. Goodwin (eds), *Introducing Human Geographies*. London: Arnold. pp. 114–21.

Goss, J. (1999b) 'Once-upon-a-time in the commodity world: an unofficial guide to the mall of America', *Annals of the Association of American Geographers*, 89 (1): 45–75.

Gottdiener, M. (2000) 'Approaches to consumption: classical and contemporary perspectives', in M. Gottdiener (ed.), *New Forms of Consumption*. Oxford: Rowman & Littlefield. pp. 3–31.

Graham, S. (1998a) 'The end of geography or the explosion of place? Conceptualizing space, place and information technology', *Progress in Human Geography*, 22 (2): 165–85.

Graham, S. (1998b) 'The spaces of surveillant-simulation: new technologies, digital representations and material geographies', *Environment and Planning D: Society and Space*, 16: 483–503.

Graham, S. (1999) 'Geographies of surveillant simulation', in M. Crang, P. Crang and J. May (eds), *Virtual Geographies: Bodies, Space and Relations*. London: Routledge. pp. 131–48.

Gramsci, A. (1971) *Selections from the Prison Notebooks*. London: Lawrence & Wishart.

Greenaway, A., Larner, W. and Le Heron, R. (2002) 'Reconstituting motherhood: milk powder marketing in Sri Lanka', *Environment and Planning D: Society and Space*, 20 (6): 719–36.

Gregory, D. (2000) 'Production of space', in D. Gregory, R.J. Johnston, G. Pratt, D. Smith and M. Watts (eds), *Dictionary of Human Geography*, 4th edn. Oxford: Blackwell. pp. 644–7.

Gregson, N. (1994) 'Beyond the high street and the mall: car boot fairs and the new geographies of consumption in the 1990s', *Area*, 26 (3): 261–7.

Gregson, N. (1995) 'And now it's all consumption?', *Progress in Human Geography*, 19 (1): 135–41.

Gregson, N. and Crewe, L. (1997a) 'The bargain, the knowledge, and the spectacle: making sense of consumption in the space of the car-boot sale', *Environment and Planning D: Society and Space*, 15: 87–112.

Gregson, N. and Crewe, L. (1997b) 'Performance and possession: rethinking the act of purchase in the light of the car boot sale', *Journal of Material Culture*, 2 (2): 241–63.

Gregson, N. and Rose, G. (2000) 'Taking Butler elsewhere: performativities, spatialities and subjectivities', *Environment and Planning D: Society and Space*, 18 (4): 433–52.

Gregson, N., Brooks, K. and Crewe, L. (2000) 'Narratives of consumption and the body in the space of the charity shop', in P. Jackson, M. Lowe, D. Miller and F. Mort (eds), *Commercial Cultures: Economies, Practices, Spaces*. Oxford: Berg. pp. 101–21.

Gregson, N., Brooks, K. and Crewe, L. (2001a) 'Bjorn again? Rethinking 70s revivalism through the reappropriation of 70s clothing', *Fashion Theory*, 5 (1): 3–28.

Gregson, N., Simonsen, K. and Vaiou, D. (2001b) 'Whose economy for whose culture? Moving beyond oppositional talk in European debate about economy and culture', *Antipode*, 33 (4): 616–46.

Gregson, N., Crewe, L. and Brooks, K. (2002a) 'Shopping, space, and practice', *Environment and Planning D: Society and Space*, 20: 597–617.

Gregson, N., Crewe, L. and Brooks, K. (2002b) 'Discourse, displacement, and retail practice: some pointers from the charity retail project', *Environment and Planning A*, 34 (9): 1661–83.

Grosz, E. (1994) *Volatile Bodies: Toward a Corporeal Feminism*. Bloomington, IN: Indiana University Press.

Hale, A. and Shaw, L.M. (2001) 'Women workers and the promise of ethical trade in the globalised garment industry: a serious beginning?', *Antipode*, 33: 510–30.

Hall, C.M. (1998) 'Making the Pacific: globalization, modernity and myth', in G. Ringer (ed.), *Destinations: Cultural Landscapes of Tourism*. London: Routledge. pp. 140–53.

Hansen, K.T. (2000) *Salaula: the World of Secondhand Clothing and Zambia*. Chicago, IL: University of Chicago Press.

Haraway, D. (1991) *Simians, Cyborgs and Women: the Reinvention of Nature*. London: Free Association.

Hartwick, E. (1998) 'Geographies of consumption: a commodity-chain approach', *Environment and Planning D: Society and Space*, 16 (4): 423–37.

Hartwick, E.R. (2000) 'Towards a geographical politics of consumption', *Environment and Planning A*, 32 (7): 1177–92.

Harvey, D. (1982) *The Limits to Capital*. Oxford: Blackwell.

Harvey, D. (1989) *The Condition of Postmodernity*. Oxford: Blackwell.
Harvey, D. (1998) 'The body as an accumulation strategy', *Environment and Planning D: Society and Space*, 16: 401–21.
Harvey, D. (2000) *Spaces of Hope*. Berkeley, CA: University of California Press.
Hay, I. (1998) 'Making moral imaginations: research ethics, pedagogy, and professional human geography', *Ethics, Place and Environment*, 1: 55–75.
Hebdige, D. (1979) *Subculture: the Meaning of Style*. London: Methuen.
Hedman, E.-L.E. and Sidel, J.T. (2000) 'Malling Manila', in *Philippine Politics and Society in the Twentieth Century*. London: Routledge. pp. 118–39.
Hetherington, K. (2004) 'Secondhandedness: consumption, disposal and absent presence', *Environment and Planning D: Society and Space*, 22 (1): 157–73.
Heyman, J.M. (2001) 'Working for beans and refrigerators: learning about environmental policy from Mexican northern-border consumers', in M.J. Cohen and J. Murphy (eds), *Exploring Sustainable Consumption: Environmental Policy and the Social Sciences*. London: Pergamon. pp. 137–55.
Hitchings, R. (2003) 'People, plants and performance: on actor network theory and the material pleasures of the private garden', *Social & Cultural Geography*, 4 (1): 99–114.
Hobson, K. (2003) 'Consumption, environmental sustainability and human geography in Australia: a missing research agenda?', *Australian Geographical Studies*, 41 (2): 148–55.
Hochschild, A.R. (2003) *The Managed Heart: Commercialization of Human Feeling*, 20th anniversary edn. Berkeley, CA: University of California.
Hollander, G.M. (2003) 'Re-naturalizing sugar: narratives of place, production and consumption', *Social & Cultural Geography*, 4 (1): 59–74.
Holliday, R. and Hassard, J. (2001) 'Contested bodies: an introduction', in R. Holliday and J. Hassard (eds), *Contested Bodies*. London: Routledge. pp. 1–17.
Holloway, L. (2002) 'Virtual vegetables and adopted sheep: ethical relations, authenticity and Internet-mediated food production technologies', *Area*, 34 (1): 70–81.
Holloway, S.L. and Valentine, G. (2001a) 'Placing cyberspace: processes of Americanization in British children's use of the Internet', *Area*, 33 (2): 153–60.
Holloway, S.L. and Valentine, G. (2001b) '"It's only as stupid as you are": children's and adults' negotiation of ICT competence at home and at school', *Social & Cultural Geography*, 2 (1): 25–42.
Hopkins, J.S.P. (1990) 'West Edmonton Mall: landscape of myths and elsewhereness', *The Canadian Geographer*, 34 (1): 2–17.
Hopkins, J.S.P. (1991) 'West Edmonton Mall as a centre for social interaction', *The Canadian Geographer*, 35 (3): 268–79.
Howes, D. (1996) 'Introduction: commodities and cultural borders', in D. Howe (ed.), *Cross-Cultural Consumption: Global Markets, Local Realities*. London: Routledge. pp. 1–18.
Hughes, A. (2000) 'Retailers, knowledges and changing commodity networks: the case of the cut flower trade', *Geoforum*, 31: 175–90.
Hughes, A. (2001) 'Global commodity networks, ethical trade and governmentality: organizing business responsibility in the Kenyan cut flower industry', *Transactions of the Institute of British Geographers*, NS 26: 390–406.
Hughes, A. and Reimer, S. (2004) 'Introduction', in A. Hughes and S. Reimer, S. (eds), *Geographies of Commodity Chains*. London: Routledge. pp. 1–16.
Hughes, G. (1998) 'Tourism and the semiological realization of space', in G. Ringer (ed.), *Destinations: Cultural Landscapes of Tourism*. London: Routledge. pp. 17–32.
Hurwitz, R. (1999) 'Who needs politics? Who needs people? The ironies of democracy in cyberspace', *Contemporary Sociology*, 28 (6): 655–61.

IISD/United Nations Department of Economic and Social Affairs (1999) *Instruments for Change: Making Production and Consumption More Sustainable*. New York: International Institute for Sustainable Development and United Nations Department of Economic and Social Affairs.

Ingham, J., Purvis, M. and Clarke, D.B. (1999) 'Hearing places, making spaces: sonorous geographies, ephemeral rhythms, and the Blackburn warehouse parties', *Environment and Planning D: Society and Space*, 17 (3): 283–305.

Jackson, P. (1989) *Maps of Meaning: an Introduction to Cultural Geography*. London: Unwin Hyman.

Jackson, P. (1993) 'Towards a cultural politics of consumption', in J. Bird, B. Curtis, T. Putnam, G. Robertson and L. Tickner (eds), *Mapping the Futures*. London: Routledge. pp. 207–28.

Jackson, P. (1999) 'Commodity cultures: the traffic in things', *Transactions of the Institute of British Geographers*, 24: 95–109.

Jackson, P. (2000) 'Cultural politics', in D. Gregory, R.J. Johnston, G. Pratt, D. Smith and M. Watts (eds), *Dictionary of Human Geography*, 4th edn. Oxford: Blackwell. p. 141.

Jackson, P. (2002a) 'Ambivalent spaces and cultures of resistance', *Antipode*, 34 (2): 326–9.

Jackson, P. (2002b) 'Commercial cultures: transcending the cultural and the economic', *Progress in Human Geography*, 26 (1): 3–18.

Jackson, P. (2002c) 'Consumption in a globalizing world', in P.J. Taylor, M.J. Watts and R.J. Johnston (eds), *Geographies of Global Change: Remapping the World*. Malden, MA: Blackwell. pp. 283–95.

Jackson, P. and Holbrook, B. (1995) 'Multiple meanings: shopping and the cultural politics of identity', *Environment and Planning A*, 27 (12): 1913–30.

Jackson, P. and Taylor, N. (1996) 'Geography and the cultural politics of advertising', *Progress in Human Geography*, 20 (3): 356–71.

Jackson, P. and Thrift, N.J. (1995) 'Geographies of consumption', in D. Miller (ed.), *Acknowledging Consumption: a Review of New Studies*. London: Routledge. pp. 204–37.

Jackson, P., Lowe, M., Miller, D. and Mort, F. (eds) (2000) *Commercial Cultures: Economies, Practices, Spaces*. Oxford: Berg.

James, J. (2000) *Consumption, Globalization and Development*. Basingstoke: Macmillan.

Johns, R. and Vural, L. (2000) 'Class, geography, and the consumerist turn: UNITE and the Stop Sweatshops Campaign', *Environment and Planning A*, 32 (7): 1193–213.

Johnston, L. and Valentine, G. (1995) 'Wherever I lay my girlfriend that's my home: performance and surveillance of lesbian identity in home environments', in D. Bell and G. Valentine (eds), *Mapping Desire: Geographies of Sexualities*. London: Routledge. pp. 99–113.

Jordan, T. (2001) 'Language and libertarianism: the politics of cyberculture and the culture of cyberpolitics', *The Sociological Review*, 49 (1): 1–17.

Kapucsinski, R. (1996) 'A normal life', *Time*, 27 May: 54–7.

Kearns, R.A. and Barnett, J.R. (1997) 'Consumerist ideology and the symbolic landscapes of private medicine', *Health and Place*, 3 (3): 171–80.

Kearns, R.A. and Barnett, J.R. (2000) '"Happy Meals" in the Starship Enterprise: interpreting a moral geography of health care consumption', *Health and Place*, 6: 81–93.

Kitchin, R.M. (1998) 'Towards geographies of cyberspace', *Progress in Human Geography*, 22 (3): 385–406.

Klein, N. (2000) *No Space, No Choice, No Jobs, No Logo: Taking Aim at the Brand Bullies*. New York: Picador.

Kollantai, V. (1999) 'Social transformations in Russia', *International Social Science Journal, UNESCO*, 159: 103–22.

Kong, L. (1995) 'Popular music in geographical analyses', *Progress in Human Geography*, 19 (2): 183–98.

Kopytoff, I. (1986) 'The cultural biography of things: commoditization as process', in A. Appadurai (ed.), *The Social Life of Things: Commodities in Cultural Perspective*. Cambridge: Cambridge University Press. pp. 64–91.

Korzeniewicz, M. (1994) 'Commodity chains and marketing strategies: Nike and the global athletic footwear industry', in G. Gereffi and M. Korzeniewicz (eds), *Commodity Chains and Global Capitalism*. Westport, CT: Greenwood. pp. 247–65.

Laermans, R. (1993) 'Learning to consume: early department stores and the shaping of the modern consumer culture (1860–1914)', *Theory, Culture and Society*, 10 (4): 79–102.

Lakoff, G. and Johnson, M. (1980) *Metaphors We Live By*. Chicago, IL: University of Chicago Press.

Larimer, T. (1998) 'Sneaker gulag: are Asian workers really exploited?', *Time*, 11 May, at http://cgi.pathfinder.com/time/asia/magazine/1998/980511/labor.html.

Larner, W. (1998) 'Hitching a ride on the tiger's back: globalisation and spatial imaginaries in New Zealand', *Environment and Planning D: Society and Space*, 16 (5): 599–614.

Larner, W. and Le Heron, R. (2002a) 'Editorial. From economic globalisation to globalising economic processes: towards post-structural political economies', *Geoforum*, 33: 415–19.

Larner, W. and Le Heron, R. (2002b) 'The spaces and subjects of a globalising economy: a situated exploration of method', *Environment and Planning D: Society and Space*, 20 (6): 753–74.

Lash, S. and Urry, J. (1987) *The End of Organized Capitalism*. Cambridge: Blackwell.

Latour, B. (1993) *We Have Never Been Modern*. New York: Harvester Wheatsheaf.

Latour, B. (1999) 'On recalling ANT', in J. Law and J. Hassard (eds), *Actor Network Theory and After*. Oxford: Blackwell. pp. 15–25.

Latour, B. and Woolgar, S. (1986) *Laboratory Life: the Construction of Scientific Facts*. Princeton, NJ: Princeton University Press.

Law, J. (1994) *Organizing Modernity*. Oxford: Blackwell.

Law, J. and Hassard, J. (eds) (1999) *Actor Network Theory and After*. Oxford: Blackwell.

Laws, G. (1995) 'Embodiment and emplacement: identities, representation and landscape in Sun City Retirement Communities', *International Journal of Aging and Human Development*, 40 (4): 253–80.

Le Heron, R.B. (2003) Personal Communication with Professor R.B. Le Heron, School of Geography and Environmental Science, University of Auckland, New Zealand.

Lee, M.J. (1993) *Consumer Culture Reborn: the Cultural Politics of Consumption*. London: Routledge.

Lees, L. (2003) 'Super-gentrification: the case of Brooklyn Heights, New York City', *Urban Studies*, 40 (12): 2487–509.

Lefebvre, H. (1991) *The Production of Space*. Oxford: Blackwell.

Leslie, D. (2002) 'Gender, retail employment and the clothing commodity chain', *Gender, Place and Culture*, 9 (1): 61–76.

Leslie, D. and Reimer, S. (1999) 'Spatializing commodity chains', *Progress in Human Geography*, 23 (3): 401–20.

Leslie, D. and Reimer, S. (2003) 'Gender, modern design, and home consumption', *Environment and Planning D: Society and Space*, 21 (3): 293–316.

Lewis, N., Moran, W., Perrier-Cornet, P. and Barker, J. (2002) 'Territoriality, enterprise and reglementation in industry governance', *Progress in Human Geography*, 26 (4): 433–62.

Leyshon, A. (2001) 'Time–space (and digital) compression: software formats, musical networks, and the reorganisation of the music industry', *Environment and Planning A*, 33: 49–77.

Lockie, S. and Kitto, S. (2000) 'Beyond the farm gate: production–consumption networks and agri-food research', *Sociologia Ruralis*, 40 (1): 3–19.

Lodziak, C. (2000) 'On explaining consumption', *Capital and Class*, 72 (Autumn): 111–33.

Longhurst, R. (1997) '(Dis)embodied geographies', *Progress in Human Geography*, 21 (4): 486–501.

Longhurst, R. (2001) *Bodies: Exploring Fluid Boundaries*. London: Routledge.

Low, N. and Gleeson, B. (1998) *Justice, Society and Nature: an Exploration of Political Ecology*. London: Routledge.

L'Orange Fürst, E. (1997) 'Cooking and femininity', *Women's Studies International Forum*, 20 (3): 441–9.

Lunt, P.K. and Livingstone, S.M. (1992) *Mass Consumption and Personal Identity: Everyday Economic Experience*. Buckingham: Open University Press.

Lury, C. (1999) 'Making time with Nike: the illusion of the durable', *Public Culture*, 11 (13): 499–526.

Mackay, H. (1997) *Consumption and Everyday Life*. London: Sage.

Maffesoli, M. (1996) *The Time of the Tribes: the Decline of Individualism in Mass Society*. London: Sage.

Mansvelt, J. (1997) 'Working at leisure: critical geographies of ageing', *Area*, 29 (4): 289–98.

Mansvelt, J. (2003) 'A choice for life? Decisions to enter retirement villages in New Zealand', in *Proceedings of the 22nd New Zealand Geographical Society Conference*, University of Auckland, pp. 219–23.

Marston, S.A. (2000) 'The social construction of scale', *Progress in Human Geography*, 24 (2): 219–42.

Marston, S.A. and Smith, N. (2001) 'States, scales and households: limits to scale thinking? A response to Brenner', *Progress in Human Geography*, 25 (4): 615–19.

Marx, K.M. (1973 [1857–8]) *The Grundrisse: Foundations of the Critique of Political Economy*. New York: Random House.

Mason, R. (1998) *The Economics of Conspicuous Consumption: Theory and Thought since 1700*. Cheltenham: Elgar.

Massey, D. (1984) *Spatial Divisions of Labour*. London: Macmillan.

Massey, D. (1993) 'Power-geometry and a progressive sense of place', in J. Bird, B. Curtis, T. Putnam, G. Robertson and L. Tickner (eds), *Mapping the Futures*. London: Routledge. pp. 59–69.

Massey, D. (1999) 'Imagining globalization: power geometries of time–space', in A. Brah, M.J. Hickman, M. Mac and Ghaill (eds), Global *Futures: Migration, Environment and Globalization*. Basingstoke: Macmillan. pp. 27–44.

May, J. (1996) 'A little taste of something more exotic: the imaginative geographies of everyday life', *Geography*, 81 (1): 57–64.

Maynard, M.L. (2001) 'Policing transnational commerce: global awareness in the margins of morality', *Journal of Business Ethics*, 30: 17–27.

McClintock, A. (1995) *Imperial Leather: Race, Gender, and Sexuality in the Colonial Conquest*. New York: Routledge.

McCormack, D. (1999) 'Body shopping: reconfiguring geographies of fitness', *Gender, Place and Culture*, 6 (2): 155–77.

McCourt, T. and Rothenbuhler, E. (1997) 'SoundScan and the consolidation of control in the popular music industry', *Media, Culture & Society*, 19 (2): 201–18.

McCracken, G. (1988) *Culture and Consumption: New Approaches to the Symbolic Character of Consumer Goods and Activities*. Bloomington, IN: Indiana University Press.

McDowell, L. (1995) 'Body work: heterosexual performances in city workplaces', in D. Bell and G. Valentine (eds), *Mapping Desire: Geographies of Sexualities*. London: Routledge. pp. 75–95.

McDowell, L. and Court, G. (1994) 'Performing work: bodily representations in merchant banks', *Environment and Planning D: Society and Space*, 12 (6): 727–50.

McGregor, H. and McMath, M. (1993) 'Leisure: a Maori and Mangaian perspective', in H. Perkins and G. Cushman (eds), *Leisure, Recreation and Tourism*. Auckland: Longman Paul. pp. 44–57.

McKendrick, N., Brewer, J. and Plumb, J.H. (1982) *The Birth of a Consumer Society: the Commercialization of Eighteenth-Century England*. Bloomington, IN: Indiana University Press.

McRobbie, A. (1993) 'Shut up and dance: youth culture and changing modes of femininity', *Cultural Studies*, 7 (3): 406–26.

McRobbie, A. (1997) 'Bridging the gap: feminism, fashion and consumption', *Feminist Review*, 55: 73–89.

McRobbie, A. (1999) *In the Culture Society: Art, Fashion and Popular Music*. London: Routledge.

Mellow, C. (1997) 'Russia's robber barons', *Time*, February 24: 54–7.

Mercier, K. (2003) 'The anti-globalisation movement in London: a coherent new social movement?' MA Thesis, Institute of Geography, Victoria University, Wellington.

Mesure, S. (2001) 'eBay tramples over Amazon'. At http://80-io.knowledge-basket.co.nz/ezproxy.massey.ac.nz/iodnews/cma/cma.pl?id=5, 31 March 2004.

Miles, S. (1998a) *Consumerism as a Way of Life*. London: Sage.

Miles, S. (1998b) 'The consuming paradox: a new research agenda for urban consumption', *Urban Studies*, 35 (5/6): 1001–8.

Miller, D. (1987) *Material Culture and Mass Consumption*. Oxford: Blackwell.

Miller, D. (ed.) (1995) *Acknowledging Consumption: a Review of New Studies*. London: Routledge.

Miller, D. (1997) *Capitalism: an Ethnographic Approach*. Oxford: Berg.

Miller, D. (1998) 'Coca-cola: a black sweet drink from Trinidad', in D. Miller (ed.), *Material Cultures: Why Some Things Matter*. London: UCL Press. pp. 169–87.

Miller, D. (2000) 'Introduction: the birth of value', in P. Jackson, M. Lowe, D. Miller and F. Mort (eds), *Commercial Cultures*. Oxford and New York: Berg. pp. 77–83.

Miller, D. (2001a) 'Behind closed doors', in D. Miller (ed.), *Home Possessions: Material Culture Behind Closed Doors*. Oxford: Berg. pp. 1–19.

Miller, D. (2001b) 'Possessions', in D. Miller (ed.), *Home Possessions: Material Culture Behind Closed Doors*. Oxford: Berg. pp. 107–21.

Miller, D. (2001c) 'The poverty of morality', *Journal of Consumer Culture*, 1 (2): 225–43.

Miller, D. (2003) 'Could the Internet defetishise the commodity?', *Environment and Planning D. Society and Space*, 21 (3): 359–72.

Miller, D. and Slater, D. (2000) *The Internet: an Ethnographic Approach*. Oxford: Berg.

Miller, D., Jackson, P., Thrift, N., Holbrook, B. and Rowlands, M. (1998) *Shopping, Place and Identity*. London: Routledge.

Mintz, S.W. (1993) 'The changing roles of food in the study of consumption', in R. Brewer and R. Porter (eds), *Consumption and the World of Goods*. London: Routledge. pp. 261–73.

Mitchell, D. (2000) *Cultural Geography: a Critical Introduction*. Malden, MA: Blackwell.

Moran, W. (1993) 'Rural space as intellectual property', *Political Geography*, 12: 263–77.

Morgan, N. and Pritchard, A. (1998) *Tourism Promotion and Power: Creating Images, Creating Identities*. Chichester: Wiley.

Morris, M. (1988) 'Things to do with shopping centres', in S. Sheridan (ed.), *Crafts: Feminist Cultural Criticism*. London: Verso. pp. 193–225.

Mort, F. (1988) 'Boy's own? Masculinity, style and popular culture', in K. Chapman and J. Rutherford (eds), *Male Order: Unwrapping Masculinity*. London: Lawrence & Wishart. pp. 193–225.

Mort, F. (1995) 'Archaeologies of city life: commercial culture, masculinity, and spatial relations in 1980s London', *Environment and Planning D: Society and Space*, 13 (5): 573–90.

Mort, F. (1998) 'Cityscapes: consumption, masculinities and the mapping of London since 1950', *Urban Studies*, 35 (5/6): 889–907.

Mort, F. (2000) 'Introduction. Paths to mass consumption: historical perspectives', in P. Jackson, M. Lowe, D. Miller and F. Mort (eds), *Commercial Cultures, Economies, Practices and Spaces*. Oxford: Berg. pp. 7–13.

Moscow Times (2001) 'IMF sees 4% growth', *Moscow Times*, 20 July, p. 6.

Murdoch, J. (1997a) 'Inhuman/nonhuman/human: actor-network theory and the prospects for a nondualistic and symmetrical perspective on nature and society', *Environment and Planning D: Society and Space*, 15 (6): 731–56.

Murdoch, J. (1997b) 'Towards a geography of heterogeneous associations', *Progress in Human Geography*, 21 (3): 321–37.

Murdoch, J. (1998) 'The spaces of actor-network theory', *Geoforum*, 29 (4): 357–74.

Murray, D. (2000) '*Haka* fracas? The dialectics of identity in discussions of a contemporary Maori dance', *The Australian Journal of Anthropology*, 11 (3): 345–57.

Nast, H. and Pile, S. (eds) (1998) *Places through the Body*. London: Routledge.

Nava, M. (1997) 'Modernity's disavowal: women, the city and the department store', in P. Falk and C. Campbell (eds), *The Shopping Experience*. London: Sage. pp. 56–92.

Neal, M.A. (1997) 'Sold out on soul: the corporate annexation of black popular music', *Popular Music and Society*, 21 (3): 117.

Nederveen Pieterse, J. (1995) 'Globalization as hybridization', in M. Featherstone, S. Lash and R. Robertson (eds), *Global Modernities*. London: Sage. pp. 45–68.

Negus, K. (1999) 'The music business and rap: between the street and the executive suite', *Cultural Studies*, 13 (3): 488–508.

Nelan, B. (1991) 'Desperate moves', *Time*, 2 September: 16–20.

Nelson, L. (1999) 'Bodies (and spaces) do matter: the limits of performativity', *Gender, Place and Culture*, 6 (4): 331–53.

Ni, C.C. and Zhang, X. (2004) 'Choking on America's e-trash', *The Dominion Post*, 14 April, B8.

Nike Incorporated (2003) *Nike Incorporated Annual Report 2003*. Nike Incorporated, http://www.nike.com/nikebiz, 25 August.

Norberg-Hodge, H. (1999a) 'Globalising resistance: turning the globalisation tide', *The Ecologist*, 29 (2): 200–6.

Norberg-Hodge, H. (1999b) 'We are all losers in the global casino: the march of the monoculture', *The Ecologist*, 29 (2): 194–7.

Olsson, G. (1980) *Birds in Egg*. London: Pion.

Osella, F. and Osella, C. (1999) 'From transience to immanence: consumption, life-cycle and social mobility in Kerala, South India', *Modern Asian Studies*, 33 (4): 989–1020.

Oushakine, S.A. (2000) 'The quantity of style: imaginary consumption in the New Russia', *Theory, Culture and Society: Explorations in Critical Social Science*, 17 (5): 97–120.

Oxfam Community Aid Abroad (2003) *Just Stop It!*, http://www.caa_org.au/compaigns/nike/faq.html, 25 August.

Pacione, M. (1997) 'Local exchange trading systems as a response to the globalisation of capitalism', *Urban Studies*, 34 (8): 1179–221.

Pacione, M. (2001) *Urban Geography: a Global Perspective*. London: Routledge.

Pain, R., Mowl, G. and Talbot, C. (2000) 'Difference and the negotiation of "old age"', *Environment and Planning D: Society and Space*, 18: 377–93.

Paolucci, G. (2001) 'The city's continuous cycle of consumption: towards a new definition of the power over time?' *Antipode*, 33 (4): 647–59.

Pawson, E. (1996) 'Landscapes of consumption', in R. Le Heron and E. Pawson (eds), *Changing Places: New Zealand in the Nineties*. Auckland: Longman Paul. pp. 318–46.

Perkins, H.C. and Thorns, D.C. (1999) 'House and home and their interaction with changes in New Zealand's urban system, households and family structures', *Housing Theory and Society*, 16: 124–35.

Perkins, H.C. and Thorns, D.C. (2001) 'Houses, homes and New Zealanders' everyday lives', in C. Bell (ed.), *The Sociology of Everyday Life in New Zealand*. Palmerston North: Dunmore. pp. 30–51.

Phillips, J. (1996) *A Man's Country? The Image of the Pakeha Male: A History*. Auckland: Penguin.

Pile, S. (1996) *The Body and the City: Psychoanalysis, Space and Subjectivity*. London: Routledge.

Plumb, J.H. (1982) 'Part III: Commercialization and society', in N. McKendrick, J. Brewer and J.H. Plumb (eds), *The Birth of a Consumer Society: the Commercialization of Eighteenth-Century England*. London: Hutchinson. pp. 265–334.

Popke, E.J. (2003) 'Poststructuralist ethics: subjectivity, responsibility and the space of community', *Progress in Human Geography*, 27 (3): 398–416.

Pratt, G. (2000) 'Research performances', *Environment and Planning D: Society and Space*, 18 (5): 639–51.

Pred, A. (1996) 'Interfusions: consumption, identity and the practices and power relations of everyday life', *Environment and Planning A*, 28 (1): 11–24.

Preteceille, E. (1986) 'Collective consumption, urban segregation, and social classes', *Environment and Planning D: Society and Space*, 4: 145–54.

Pritchard, B. (2000) 'The transnational corporate networks of breakfast cereals in Asia', *Environment and Planning A*, 32: 789–804.

Pritchard, W.N. (2000) 'Beyond the modern supermarket: geographical approaches to the analysis of contemporary Australian retail restructuring', *Australian Geographical Studies*, 38 (2): 204–18.

Proctor, J.D. and Smith, D.M. (1999) *Geography and Ethics: Journeys in a Moral Terrain*. London: Routledge.

Purcell, M. (2003) 'Islands of practice and the Marston/Brenner debate: toward a more synthetic critical human geography', *Progress in Human Geography*, 27 (3): 317–32.

Purdue, D., Durrschmidt, J., Jowers, P. and O'Doherty, R. (1997) 'DIY culture and extended milieux: LETS, veggie boxes and festivals', *The Sociological Review*, 45 (4): 645–67.

Purvis, M. (1998) 'Societies of consumers and consumer societies: co-operation, consumption and politics in Britain and continental Europe c. 1850–1920', *Journal of Historical Geography*, 24 (2): 147–69.

Purvis, M. (2003) 'Societies of consumers and consumer societies', in D.B. Clarke, M.A. Doel and K.M.L. Housiaux (eds), *The Consumption Reader*. London: Routledge. pp. 69–76.

Quirke, M. (2002a) 'Nothing derogatory says BBC', *The Dominion*, 13 April: 3.

Quirke, M. (2002b) 'This is the BBC: ka mate, ka mate!', *The Dominion*, 6 April: 1.

Qureshi, K. and Moores, S. (1999) 'Identity remix: tradition and translation in the lives of young Pakistani Scots', *European Journal of Cultural Studies*, 2 (3): 311–30.

Raikes, P., Jensen, M.F. and Ponte, S. (2000) 'Global commodity chain analysis and the French *filère* approach: comparison and critique', *Economy and Society*, 29 (3): 390–417.

Ramírez, B. (2000) 'Guest editorial. The politics of constructing an international group of critical geographers and a common space of action', *Environment and Planning D: Society and Space*, 18 (5): 537–43.

Rappaport, E.D. (2000) *Shopping for Pleasure*. Chichester: Princeton University Press.

Redclift, M. (1996) *Wasted: Counting the Costs of Global Consumption*. London: Earthscan.

Redfern, P.A. (1997) 'A new look at gentrification. 1: Gentrification and domestic technologies', *Environment and Planning A*, 29 (7): 1275–96.

Renard, M.-C. (1999) 'The interstices of globalization: the example of fair trade coffee', *Sociologia Ruralis*, 39 (4): 484–500.

Revill, G. (1998) 'Samuel Coleridge-Taylor's geography of disappointment: hybridity, identity and networks of musical meaning', in A. Leyshon, D. Matless and G. Revill (eds), *The Place of Music*. London: Guilford. pp. 197–221.

Revill, G. (2000) 'Music and the politics of sound: nationalism, citizenship, and auditory space', *Environment and Planning D: Society and Space*, 18 (5): 597–613.

Ritzer, G. (1993) *The McDonaldization of Society: an Investigation into the Changing Character of Contemporary Social Life*. Newbury Park, CA: Pine Forge.

Ritzer, G. (1999) *Enchanting a Disenchanted World: Revolutionizing the Means of Consumption*. Thousand Oaks, CA: Pine Forge.

Ritzer, G. (2002) 'An introduction to Mcdonaldization', in G. Ritzer (ed.), *McDonaldization: The Reader*. London: Sage. pp. 141–7.

Ritzer, G., Goodman, D. and Wiedenhoft, W. (2000) 'Theories of consumption', in G. Ritzer and B. Smart (eds), *Handbook of Social Theory*. London: Sage. pp. 410–27.

Roberts, M.L. (1998) 'Review essay. Gender, consumption, and commodity culture', *American Historical Review*, 103 (3): 817–44.

Robson, K. (1998) '"Meat" in the machine: the centrality of the body in internet interactions', in J. Richardson and A. Shaw (eds), *The Body in Qualitative Research*. Aldershot: Ashgate. pp. 185–97.

Rodaway, P. (1995) 'Exploring the subject in hyper-reality', in S. Pile and N. Thrift (eds), *Mapping the Subject: Geographies of Cultural Transformation*. London: Routledge. pp. 241–66.

Rojek, C. (1995) *Decentering Leisure, Rethinking Leisure Theory*. London: Sage.

Rose, G. (1993) *Feminism and Geography*. Oxford: Polity.

Rose, G. (1995) 'Geography and gender, cartographies and corporealities', *Progress in Human Geography*, 19 (4): 544–48.

Rose, G. (1997) 'Situating knowledges: positionality, reflexivities and other tactics', *Progress in Human Geography*, 21: 305–20.

Rowlands, M. (1994) 'The material culture of success: ideals and life cycles in Cameroon', in J. Friedman (ed.), *Consumption and Identity*. Chur, Switzerland: Harwood. pp. 147–66.

Rule, J.B. (1999) 'Silver bullets or land rushes? Sociologies of cyberspace', *Contemporary Sociology*, 28 (6): 661–4.

Ryan, C. and Crotts, J. (1997) 'Carving and tourism: a Maori perspective', *Annals of Tourism Research*, 24: 898–918.

Sack, R.D. (1988) 'The consumer's world: place as context', *Annals of the Association of American Geographers*, 78 (4): 642–64.

Sack, R.D. (1992) *Place, Modernity, and the Consumer's World: a Relational Framework for Geographical Analysis*. Baltimore, MD: Johns Hopkins University Press.

Sack, R.D. (1997) *Homo Geographicus: A Framework for Action, Awareness, and Moral Concern*. Baltimore, MD: Johns Hopkins University Press.

Sack, R.D. (1999) 'A sketch of a geography theory of morality', *Annals of the Association of American Geographers*, 89 (1): 26–44.

Said, E. (1978) *Orientalism*. London: Routledge.

Samers, M. and Sidaway, J. (2000) 'Guest editorial: exclusions, inclusions, and occlusions in "Anglo-American geography": reflections on Minca's "Venetian geographical praxis"', *Environment and Planning D: Society and Space*, 18 (6): 663–6.

Saunders, P. (1989) 'The sociology of consumption: a new research agenda', in P. Otnes (ed.), *The Sociology of Consumption*. Atlantic Highlands, NJ: Humanities. pp. 141–56.

Sayer, A. (2001) 'For a critical cultural political economy', *Antipode*, 33 (4): 687–708.

Sayer, A. (2003) '(De)commodification, consumer culture, and moral economy', *Environment and Planning D: Society and Space*, 21: 341–57.

Scanlon, J. (ed.) (2000) *The Gender and Consumer Culture Reader*. New York: New York University Press.

Scheyvens, R. (2002) *Tourism for Development: Empowering Communities*. Harlow: Prentice Hall.

Schoenberger, E. (1998) 'Discourse and practice in human geography', *Progress in Human Geography*, 22 (1): 1–14.

Schofield, J. (2001) 'Advertisers aim below the line for customers', *Moscow Times*, p. 7.

Scott, A.J. (2000) *The Cultural Economy of Cities*. London: Sage.

Sell, B. (1999) '"Plastic poi" culture seen as degrading', *New Zealand Herald*, 1 June.

Serres, M. and Latour, B. (1995) *Conversations of Science, Culture and Time*, trans. R. Lapidus. Ann Arbor, MI: University of Michigan Press.

Seyfang, G. (2001) 'Community currencies: small change for a green economy', *Environment and Planning A*, 33 (6): 975–96.

Shammas, C. (1993) 'Changes in English and Anglo-American consumption from 1550 to 1800', in J. Brewer and R. Porter (eds), *Consumption and the World of Goods*. London: Routledge. pp. 177–205.

Shapiro, S. (1998) 'Places and spaces: the historical interaction of technology, home and privacy', *The Information Society*, 14: 275–84.

Shields, R. (1989) 'Social spatialization and the built environment: the West Edmonton Mall', *Environment and Planning D: Society and Space*, 7 (2): 147–64.

Shields, R. (1992a) 'The individual, consumption cultures and the fate of community', in R. Shields (ed.), *Lifestyle Shopping: the Subject of Consumption*. London: Routledge. pp. 99–113.

Shields, R. (ed.) (1992b) *Lifestyle Shopping: the Subject of Consumption*. London: Routledge.

Short, J.R., Boniche, A., Kim, Y. and Li, P.L. (2001) 'Cultural globalization, global English, and geography journals', *Professional Geographer*, 53 (1): 1–11.

Shurmer-Smith, P. and Hannam, K. (1994) *Worlds of Desire, Realms of Power: a Cultural Geography*. London: Arnold.

Sik, E. and Wallace, C. (1999) 'The development of open-air markets in east-central Europe', *International Journal of Urban and Regional Research*, 23 (4): 697–714.

Silicon Valley Toxics Coalition (2004) 'Poison PCs and toxic TVs: e-waste tsunami to roll across the US: are we prepared?' At http://www.svtc.org/cleancc/pubs/ppcttv 2004execsum.htm, 4 May 2004.

Silvey, R. (2002) 'Sweatshops and the corporatization of the university', *Gender, Place and Culture*, 9 (2): 201–7.

Simmel, G. (1978 [1907]) *The Philosophy of Money*. London: Routledge.
Sissons, J. (1993) 'The systematisation of tradition: Maori culture as a strategic resource', *Oceania*, 64 (2): 97–109.
Skelton, T. and Valentine, G. (eds) (1998) *Cool Places: Geographies of Youth Cultures*. London: Routledge.
Slater, D. (1997) *Consumer Culture and Modernity*. Oxford: Polity.
Slater, D. (2000) 'Consumption without scarcity: exchange and normativity in an internet setting', in P. Jackson, M. Lowe, D. Miller and F. Mort (eds), *Commercial Cultures: Economies, Practices, Spaces*. Oxford: Berg. pp. 123–42.
Slocum, R. (2004) 'Consumer citizens and the cities for climate protection campaign', *Environment and Planning A*, 36: 763–82.
Smart, B. (1999) *Resisting McDonaldization*. London: Sage.
Smith, A. (2002) 'Culture/economy and spaces of economic practice: positioning households in post-communism', *Transactions of the Institute of British Geographers*, NS 27: 232–50.
Smith, D.M. (1997) 'Geography and ethics: a moral turn?', *Progress in Human Geography*, 21: 583–90.
Smith, D.M. (1999) 'Geography and ethics: how far should we go', *Progress in Human Geography*, 23: 119–25.
Smith, D.M. (2000) 'Social justice revisited', *Environment and Planning A*, 32 (7): 1149–62.
Smith, D.M. (2001) 'Geography and ethics: progress or more of the same?', *Progress in Human Geography*, 25: 261–8.
Smith D.W. (1998) 'Urban food systems and the poor in developing countries', *Transactions of the Institute of British Geographers*, 23: 207–19.
Smith, M.D. (1996) 'The Empire filters back: consumption, production, and the politics of Starbucks coffee', *Urban Geography*, 17: 502–25.
Smith, R.G. (2003) 'World city actor-networks', *Progress in Human Geography*, 27 (1): 25–44.
Smith, S.J. (1994) 'Soundscape', *Area*, 26 (3): 232–40.
Smith, S.J. (2000) 'Performing the (sound) world', *Environment and Planning D: Society and Space*, 18 (5): 615–37.
Soja, E. (1989) *Postmodern Geographies: the Reassertion of Space in Critical Social Theory*. London and New York: Verso.
Soja, E. (1996) *Thirdspace: Journeys to Los Angeles and Other Real-and-Imagined Places*. Cambridge: Blackwell.
Sorkin, M. (1992) 'See you in Disneyland', in *Variations on a Theme Park: the New American City and the End of Public Space*. New York: Hill and Wang. pp. 205–32.
Squire, S.J. (1998) 'Rewriting languages of geography and tourism: cultural discourses of destinations, gender and tourism history in the Canadian Rockies', in G. Ringer (ed.), *Destinations: Cultural Landscapes of Tourism*. London: Routledge. pp. 80–100.
Statistics New Zealand (2001) *Statistics New Zealand*, http://www.stats.govt.nz, 10 July.
Stearns, P.N. (2001) *Consumerism in World History: the Global Transformation of Desire*. London: Routledge.
Steed, B. (2000) 'Maori culture a blend of old and new', *The Evening Post*, 19 October: 23.
Stevenson, N., Jackson, P. and Brooks, K. (2000) 'Ambivalence in men's lifestyle magazines', in P. Jackson, M. Lowe, D. Miller and M. Frank (eds), *Commercial Cultures: Economies, Practices, Spaces*. Oxford: Berg. pp. 189–212.
Suryanata, K. (2002) 'Diversified agriculture, land use, and agrofood networks in Hawaii', *Economic Geography*, 78 (1): 71–86.

Swanson, G. (1995) '"Drunk with glitter": consuming spaces and sexual geographies', in Watson & Gibson (eds), *Postmodern Cities and Spaces*. Cambridge: Blackwell. pp. 80–98.

Tan, S.B.-H. (2000) 'Coffee frontiers in the central highlands of Vietnam: networks of connectivity', *Asia Pacific Viewpoint*, 41 (1): 51–67.

Taylor, F.W. (1967 [1911]) *Principles of Scientific Management*. New York: Norton.

Taylor, J.P. (1998) *Consuming Identity: Modernity and Tourism in New Zealand*. Auckland: Department of Anthropology, University of Auckland.

The Dominion (2001) 'Pop star's wearing "my tribe's tattoo"', *The Dominion*, 5 January: 3.

The Dominion (2002) 'Mana of haka and moko diluted says scholar', *The Dominion*, 16 April: 19.

TheStaffordGroup (2001) *A Study of Barriers, Impediments and Opportunities for Maori in Tourism: he matai tapoi Maori*. Prepared for the Office of Tourism and Sport and Te Puni Kokiri. Wellington: TheStaffordGroup, Woollahra, NSW.

Thrift, N. (2000a) 'Actor network theory', in D. Gregory, R.J. Johnston, G. Pratt, D. Smith and M. Watts (eds), *Dictionary of Human Geography*, 4th edn. Oxford: Blackwell. pp. 5.

Thrift, N. (2000b) 'Afterwords', *Environment and Planning D: Society and Space*, 18 (2): 213–55.

Thrift, N. (2000c) 'Performing cultures in the new economy', *Annals of the Association of American Geographers*, 90 (4): 674–92.

Thrift, N. (2000d) 'It's the little things', in K. Dodds and D. Atkinson (eds), *Geopolitical Traditions: a Century of Geopolitical Thought*. London: Routledge. pp. 380–7.

Thrift, N. and Glennie, P. (1993) 'Historical geographies of urban life and modern consumption', in G. Kearns and C. Philo (eds), *Selling Places: the City as Cultural Capital, Past and Present*. Oxford: Pergamon. pp. 33–48.

Tikhomirov, V. (1996) 'CIS: new moves towards a closer integration', *International Law News*, 31: 38–9.

Tourism New Zealand (1995) *Product Development Opportunities for Asian Markets*. Wellington: New Zealand Tourism Board.

Trentmann, F. (2004) 'Beyond consumerism: new historical perspectives on consumption, *Journal of Contemporary History*, 39 (3): 373–401.

Turkle, S. (2002) 'Life on the screen: identity in the age of the Internet', in M.J. Dear and S. Flusty (eds), *The Spaces of Postmodernity: Readings in Human Geography*. Oxford: Blackwell. pp. 455–62.

United Nations Development Programme (1999) 'Facts and figures on poverty', *United Nations Development Programme, Sustainable Human Development*, http://www.undp.org/teams/english/facts.htm

Urry, J. (1995) *Consuming Places*. London: Routledge.

Valentine, G. (1999a) 'Consuming pleasures: food, leisure and the negotiation of sexual relations', in D. Crouch (ed.), *Leisure/Tourism Geographies*. London: Routledge. pp. 164–80.

Valentine, G. (1999b) 'A corporeal geography of consumption', *Environment and Planning D: Society and Space*, 17 (3): 329–51.

Valentine, G. (1999c) 'Eating in: home, consumption and identity', *The Sociological Review*, 47 (3): 491–524.

Valentine, G. (2002) 'In-corporations: food, bodies and organizations', *Body & Society*, 8 (2): 1–20.

Veblen, T. (1975 [1899]) *The Theory of the Leisure Class*. New York: Augustus.

REFERENCES

Wakeford, N. (1999) 'Gender and landscapes in computing', in M. Crang, P. Crang and J. May (eds), *Virtual Geographies: Bodies, Spaces and their Relations*. London: Routledge. pp. 178–201.

Wallace, C. and Kovacheva, S. (1996) 'Youth cultures and consumption in Eastern and Western Europe', *Youth & Society*, 28 (2): 189–214.

Wallerstein, I: (1974) 'The rise and future demise of the world capitalist system: concepts for comparative analysis', *Comparative Studies in Society and History*, 16 (4): 387–415.

Wallerstein, I. (1983) *Historical Capitalism*. London: Verso.

Walmsley, D.J. (2000) 'Community, place and cyberspace', *Australian Geographer*, 31 (1): 5–19.

Ward, G. (1997) *Postmodernism*. London: Hodder Headline.

Warde, A. (1992) 'Notes on the relationship between production and consumption', in R. Burrows and C. Marsh (eds), *Consumption and Class*. London: Macmillan. pp. 15–31.

Warf, B. (1994) 'Geographical review: *Place, Modernity and the Consumer's World*, by Robert David Sack', *Annals of the Association of American Geographers*, 84: 106–9.

Warren, S. (1999) 'Cultural contestation at Disneyland Paris', in D. Crouch (ed.), *Leisure/Tourism Geographies: Practices and Geographical Knowledge*. London: Routledge. pp. 109–25.

Watson, J.L. (1997) *Golden Arches East*. Stanford, CA: Stanford University Press.

Watts, M. and Watts, N. (1983) *Silent Violence: Food, Famine and Peasantry in Northern Nigeria*. Berkeley, CA: University of California Press.

Watts, M.J. (1999) 'Commodities', in P. Cloke, P. Crang and M. Goodwin (eds), *Introducing Human Geographies*. London: Arnold. pp. 305–15.

Webster, S. (1993) 'Postmodernist theory and the sublimation of Maori culture', *Oceania*, 63 (3): 222–33.

Whatmore, S. (2003) 'Editorial. From banana wars to black Sigatoka: another case for a more-than-human geography', *Geoforum*, 34: 139.

Whatmore, S. and Thorne, L. (1997) 'Nourishing networks: alternative geographies of food', in D. Goodman and M.J. Watts (eds), *Globalising Food: Agrarian Questions and Global Restructuring*. London: Routledge. pp. 287–304.

Wilk, R. (1995) 'Learning to be local in Belize: global systems of common difference', in D. Miller (ed.), *Worlds Apart: Modernity through the Prism of the Local*. London: Routledge. pp. 110–33.

Wilk, R. (2001) 'Consuming morality', *Journal of Consumer Culture*, 1 (2): 245–60.

Wilk, R. (2002) 'Consumption, human needs, and global environmental change', *Global Environmental Change*, 12: 5–13.

Williams, C.C. (1996) 'Local exchange and trading systems: a new source of work and credit for the poor and unemployed?', *Environment and Planning A*, 28 (8): 1395–415.

Williams, P., Hubbard, P., Clark, D. and Berkeley, N. (2001) 'Consumption, exclusion and emotion: the social geographies of shopping', *Social & Cultural Geography*, 2 (2): 203–20.

Winchester, H. (1992) 'The construction and deconstruction of women's roles in the urban landscape', in K. Anderson & F. Gale (eds), *Inventing Places: Studies in Cultural Geography*. Melbourne: Longman Cheshire. pp. 139–56.

Winship, J. (2000) 'Culture of restraint: the British chain store 1920–1939', in P. Jackson, M. Lowe, D. Miller and F. Mort (eds), *Commercial Cultures: Economies, Practices, Spaces*. Oxford: Berg. pp. 15–34.

Woodward, I., Emmison, M. and Smith, P. (2000) 'Consumerism, disorientation and postmodern space: a modest test of immodest theory', *British Journal of Sociology*, 51 (2): 339–54.

Wrigley, N. and Lowe, M. (eds) (1996) *Retailing, Consumption and Capital: Towards the New Retail Geography*. Harlow: Longman.

Wrigley, N. and Lowe, M. (eds) (2002) *Reading Retail*. New York: Oxford University Press.

Wrigley, N., Lowe, M. and Currah, A. (2002) 'Retailing and e-tailing', *Urban Geography*, 23 (2): 180–97.

Wrigley, N., Warm, D. and Margetts, B. (2003) 'Deprivation, diet, and food-retail access: findings from the Leeds "food deserts" study', *Environment and Planning A*, 35 (1): 151–88.

Yasmeen, G. (1995) 'Exploring a foodscape: the case of Bangkok', *Malaysian Journal of Tropical Geography*, 26 (1): 1–11.

Zukin, S. (1998) 'Urban lifestyles: diversity and standardisation in spaces of consumption', *Urban Studies*, 35 (5/6): 825–39.

Index

Abaza, M. 65
Abelson, E. 41
actants 117, 120, 121, 123
active retirees 87–8
actor network theory (ANT) 19, 117–23, 124
  food production 18
  moralities 155
  performativity 149
Adorno, T. 24
advertising 37
  commodity racism 38–9
  consumer's world 58
  Russia 51
  social consciousness 42
affinity 84
ageing bodies 21, 86–8
Agenda 21 158
alternative economic spaces 15–16
alternative geographies 43
alternative sites of consumption 66–9
alternative spaces of consumption 14–15
Anokhi 139
Appadurai, Arjun 7–8
appropriation 144, 145
arcades project 60
Argenbright, R. 50
Arnold, Barbara 162–4
authenticity 143, 144
Azaryhau, Maoz 136, 137

backstage 89–90, 92
Banim, M. 85
Barnett, Ross 136–7
Baudrillard, Jean 14, 44, 46, 81
BBC 143–4
Beaverstock, J.V. 157
Belcher, Sophie 130
Bell, D. 85, 94
Benjamin, Walter 60
Bennett, A. 133
Bentham, J. 26
Bhabha, H. 138
Bilkent shopping centre 65–6

Binnie, J. 82, 84
black box of consumption 121–2
Blomley, N. 40
body 21, 22, 84–93, 95–6
  art 86
  journeys 99
Bourdieu, P. 82–3, 154, 156
Bowler, S.M. 63
branding 37
Bread Alert exercise 162–3
Brooks, Kate 67, 92–3
Bryman, A. 128
Burt, S. 109
Butler, Judith 85, 90–1

Cafédirect 120–1
Cahill, S. 86
Callon, Michel 117
Cameron, J. 73
Campbell, H. 116
capitalism 25, 32
  Fordism 42
  postmodernity 44
  Russia 49–52
car boot sales 14, 66, 67–8, 91
carnivalesque 93, 94
Carroll, J. 129
charity shops 67–8, 92
Cities for Climate Protection (CCP) 159
Clancy, M. 109
class system 82, 83
  emulation 36–7
  shoplifting 40, 41, 155
climate protection 159
clothing industry 22, 92–3, 94, 95, 111–14, 139
coffee consumption 120–2
commercial cultures 19–20, 127–47
commodification 2–3, 5, 8, 20, 23–4, 25, 35, 37, 44
  moral economy 153–4
  music 132, 133
  Russia 49–52
  shopping malls 62

commodity chains 18, 101–14, 155
  ANT 119
  global 102–10, 114, 124, 150
commodity circuits 19, 114–17, 124
commodity fetishism 23–4, 25, 38, 44, 154, 160
commodity racism 38–9
communities of consumption 46
connections 101–26
Connell, J. 129
conspicuous consumption 36–7
consumer activism 18, 105–8, 155
consumer citizens 159
consumer culture 127, 153–4
consumer pull 104
consumer revolution 32
consumer societies 43
Cook, I. 25, 96
cooperative societies 43
corporeality 84, 93, 98
Court, G. 90, 128
Crang, M. 78
Crang, P. 25, 90, 96
creolization 137–40
Crewe, Louise 67–8, 92–3
critical social science 10–11
Crompton, R. 82
Cross, G. 37
cultural capital 83, 154
cultural economies 19
cultural politics 17, 21–2, 26, 84, 113
cultural symbolic exchange 8
cyberspace 3–4, 73–7

Debord, G. 46
deconstruction 26
de-differentiation 3
DeNora, Tia 130
department stores 39–40, 60–1
disintermediation 76
Disney 59, 128

displacement 96, 99, 116
DIY 70–2, 73
domesticity 34, 70, 72
Domosh, Mona 39, 40–1
Douglas, M. 8
Du Gay, P. 114–15, 131
Dwyer, C. 139

EAST 139
eBay 3–4
e-commerce 76
economic spaces, alternative 15–16
economic value, tourism 141
economies 23, 35
  cultural 19
  culture interdependence 128
  moral 153–4
  Russia 50
  see also political economy
Edwards, T. 2, 36
embodiment 21, 88–93, 97, 98, 99, 155
emotional labour 90
emplacement 88–93, 97, 99, 155
emulation 36–7
Enlightenment 35
environment
  Agenda 21 158
  climate protection 159
  global change 156–7
e-retailer 76–7
Erkip, Feyzan 65–6
ethics 17, 18
ethnography 14, 73
  arcades 60
  Internet 76
  shopping malls 63–4
  socialities 17
e-waste 4–5
exchange 7, 17
  cultural symbolic 8
  LETS 15–16
  rituals 9
exports, e-waste 5

fair-trade coffee 120–1
fat bodies 21, 86
Featherstone, M. 48
femininity 86
feminism 17, 84
filière 110
Fine, Ben 19, 27, 110, 111, 114, 116, 124
Fiske, J. 66
flâneur 60, 62, 65

food
  deserts 152
  ethical relations 18
  feminism 17
  foodscapes 97
  geographies 94–9
  networks 120
  politics 150
Fordism 42, 103
Foucault, M. 26, 159
frontstage 89–90, 92

Gelber, S.M. 70
gender
  department stores 39–40, 60–1
  femininity 86
  food 98
  home 34, 69, 70–3
  identity 84, 90–1
  leisure class 36
  masculinity 22, 70–3, 86
  overconsumption 40–1
  wardrobe moment 85
Gereffi, G. 103–5
Girl Guides 162–4
Gleeson, B. 156
Glennie, Paul 19, 33–4, 37, 48–9, 63
global commodity chains 102–10, 114, 124, 150
global environmental change 156–7
global homogenization 134–7
globalization 66
  commercial cultures 127, 134–40
  commodity chains 105
  as homogenization 6
  space 56–7
Goffman, Erving 89–90, 92, 130
Golani Junction 136, 137
Goodman, D. 122
Goss, John 62–3, 64, 81
Graham, S. 73
Gramsci, Antonio 24
Greenaway, Alison 123
Gregson, Nicky 10, 22, 63–4, 67–8, 90–1, 92–3, 128
Grosz, Elizabeth 85

habitus 83, 84, 156–7
haka 144, 145
Hansen, Karen Tranberg 111–14
Hartwick, E. 107, 108

Harvey, D. 44, 56, 99, 160
hegemony 24, 43
Heinz Corporation 39
heterosexuality 69–70
hip-hop music 132–3
Hitchings, R. 119, 148
Hobson, Kersty 158
Hochschild, A.R. 90
Hollander, G.M. 39, 116
Holloway, Lewis 18
Holloway, S.L. 76
home 16, 17
  domesticity 34, 70, 72
  ownership 42
  public/private spheres 69–73
horizontal provision systems 19
Horkheimer, M. 24
Howes, D. 134
Hughes, A. 118–19
humanism 20
human rights issues, Nike 106–8
Hutley, Jo 144
hybridity 59, 133, 137–40, 144, 145
hyper-reality 46

identity 25–6, 80–100
  formation 80–3, 84, 85, 86, 88, 130
  value 83, 99
Imperial Leather 38–9
indigenous people 140, 141, 142–5
industrial revolution 32, 34
inequalities 151
intellectual property 144, 145
international community 158
Internet 17, 73–7
  commodity chains 105
  eBay 3–4
  ethical food relations 18
  music 130–1
  Nike protests 106–7
  postmodernity 48
Isherwood, B. 8

Jackson, P. 128, 139
Johns, R. 107, 108

Kapucsinski, R. 49–50
Kearns, Robin 136–7
Kitchin, R.M. 75
Knight, Phil 106, 108
knowingness 93, 95
Kopytoff, Igor 7–8

INDEX

landscapes
  consumer's world 58
  identities 86
  sacred 136–7
  shopping malls 64
  therapeutic 136–7
language, sign system 25
Larner, Wendy 123
Latour, Bruno 117, 118, 149
Law, John 117
Laws, Glenda 86, 87–8
Lee, M.J. 7, 42
Lefebvre, Henri 56–7
Le Heron, Richard 123
leisure 10
  active retirees 87
  lifestyle shopping 48
leisure class 36–7
Leopold, Ellen 19, 110, 124
Leslie, D. 111
Liepins, R. 116
lifestyle shopping 48, 81, 82
linearity 19
lived bodies 84
local exchange trading schemes (LETS) 15–16
Longhurst, R. 84–5
Lowe, Michelle 128
Low, N. 156
Lung, C. 106

McClintock, A. 38–9
McDonalds 50, 135–7
McDowell, L. 90, 128
Mackay, H. 81
McKendrick, N. 36, 37
McRobbie, A. 82
Maori 141, 142–5
maps of meaning 85
Margetts, Barrie 152
marketing 37
Marston, S.A. 57
Marx, Karl 20, 23–5, 42, 154
masculinity 22, 70–3, 86
Massey, Doreen 27, 56, 123
material culture 8–9
May, Jon 96
meaning 7, 8, 9, 25–6, 85, 139
Mellow, C. 50
Miles, S. 2, 161
milk consumption 123
Miller, Daniel 9, 63, 76, 83, 93, 99, 128, 153, 155
modernity 32–3, 35, 39, 49, 61
  capitalism 42
  cf. postmodernity 45
  space 56–7

moral economy 153–4
moralities 148–65
Morris, M. 64
Mort, Frank 1, 22, 77, 127, 128, 155
motherhood 123
Movement of Mothers to Combat Malnutrition (MMCM) 123
music, commercial culture 128, 129–34

Nederveen Pieterse, J. 138
Negus, K. 131, 132–3
Nelson, L. 91–2
networks 18, 116, 120
  transnational 96
  see also actor network theory
New York woman 40–1
New Zealand Dairy Board (NZDB) 123
niche consumption 44–9
Nicholas, Darcy 144
Nike 105–8
non-linear approaches 116–17, 139

objective culture 9
Omidyar, Pierre 3
organic food 116
Oushakine, S.A. 52

Pain, R. 82
panopticon 26
performance 89–90, 91–2, 130, 163–4
performativity 89, 90–2, 99, 148–52
Perkins, H.C. 69, 72
plant-human relationships 119
point-of-sale technologies 131
political economy 19, 26–7, 103, 122–3
  context 18
  Marx 23, 24–5
politics 124
  commercial culture 129–34
  cultural 17, 21–2, 26, 84, 113
  food 150
  hip-hop music 132
  moralities 160–4
  music 133–4
positionality 82
post-Fordism 46, 48
postmodernity 2, 44–9, 80–4
poststructuralism 20, 21, 23, 25–8, 117, 122–3, 154

power 23–8, 124–5
  ANT 118, 122, 123
  commodity chains 109
  Foucault 26
  geometries 27, 123, 124, 160
  home 69
  identity 84, 90, 91
  mapping masculinity 22
  moralities 160–1
  music 129, 131
  performativity 150
praxis 160–4
production 7, 16
  connections 101–26
  consumption gap 161
  food 18
  Marx 23–5
  power geometries 27
productionist bias 19
psychoanalysis 84
public spaces 39–41, 65–6, 73
Purcell, M. 149
Purvis, Martin 43
Putin, Vladimir 50

racism, commodity 38–9
Ramírez, B. 152
rap music 132–3
regulatory frameworks 18, 20
Reimer, S. 111
relational framework 58, 160
religion 35, 64
representations of space 57
reterritorialization 138–9
retirement villages 87–8
retro clothing 93, 95
retro music 46
Riley, S. 86
risk society 46
Ritzer, G. 59, 135
Romanticism 35
Rose, Gillian 90–1
Rowlands, M. 81
Russian Federation 49–52

Sack, Robert 10, 58, 160
sacred landscapes 136–7
Said, E. 140
Saunders, P. 82
Sayer, Andrew 153–4
scale 14, 56–8, 69–73, 118
Scanlon, J. 72
second-hand goods 14, 17, 22, 67–8, 92–3, 94, 95, 111–14
self
  identities 46, 80–1, 83
  performance 89

semiotics 63–4
Serres, Michel 117
sex tourism 109
sexuality 98
Sharples, Pita 144
Shields, Rob 48
shoplifting 40, 41, 155
shopping
   arcades project 60
   charity shops 67–8, 92
   department stores
      39–40, 60–1
   instore music 130
   lifestyle 48, 81, 82
   malls 61–6
Simmel, G. 9
simulation 46
Sissons, J. 141
Slater, D. 35, 76
Slocum, Rachel 159
Smith, M.D. 120, 121–2
social centrality 65
social consciousness 42
social construction 84–5, 119
social context 90
social exclusion 152
social groups 82
socialities
   of consumption 12–13,
      16–20
   Turkish shopping malls 65–6
social justice 162
social life 7–8
social relations 1, 6, 102, 124
   Internet 74, 75
   Marx 24
   SOP 110
social structures 82–3

Soja, Edward 56, 57
Solomon, Maui 144
Sony Walkman 115
spaces 56–79
   consumer's world 58
   food 97–8
   politics 162
   of representation 57, 59
   spectacular 59
   trialectic 57
Sparks, L. 109
spatial analysis 118
spatialities of consumption
      11–12, 14–16
spatial practice 57
spectacular spaces 59
Starbucks 121
state 20, 152
subjectivities 26
   of consumption 13, 20–3
   identities 93
   lifestyle shopping 48
   Turkish shopping malls 65–6
supermarkets 39, 116
surveillance 26, 63, 73, 74, 78
sustainability 158
Swanson, Gillian 84
symbolic constructs 96
symbolic exchange 8
systems of provision 19,
      109–14, 124

Taylorism 42
technology, e-waste 4–5
textual research 63–4
therapeutic landscapes 136–7
Thorne, L. 120–1
Thorns, D. C. 69, 72

three-stage model 49, 52–3
Thrift, Nigel 19, 33–4, 37,
   48–9, 63, 91–2
tourism 10, 46, 47, 75, 109,
   140–5
transcultural convergence 138
transformation 160
   placing 49–52
   Turkish shopping malls 66
transnationalism 116–17,
   137–40, 144, 145, 146
transnational networks 96
Trentmann, F. 49
Trinidad 76
Turkle, S. 77

United Nations 158

Valentine, Gill 76, 85, 94, 97–9
Veblen, Thorstein 36–7, 154
vertical provision systems 19
Vural, L. 107, 108

Walker, Charles 144
Wal-Mart 109
wardrobe moment 85
Warm, Daniel 152
Warren, Stacey 59
Watts, M. 101–2, 158–9
Watts, N. 158–9
Webster, S. 145
Whatmore, S. 120–1
Whitlams 129
Wilk, Richard 155–7, 160
world systems theory 102–3
Wrigley, Neil 76, 152

Zola, Emile 40

Indexed by Caroline Eley